MyMathLab® Notebook

CHRISTINE VERITY

Basic Mathematics

Second Edition

John Squires
Chattanooga State Community College

Karen Wyrick
Cleveland State Community College

PEARSON

Boston Columbus Indianapolis New York San Francisco
Amsterdam Cape Town Dubai London Madrid Milan Munich Paris Montréal Toronto
Delhi Mexico City São Paulo Sydney Hong Kong Seoul Singapore Taipei Tokyo

Reproduced by Pearson from electronic files supplied by the author.

Publishing as Pearson, 501 Boylston Street, Boston, MA 02116.

ISBN-13: 978-0-13-404915-1
ISBN-10: 0-13-404915-2

1 2 3 4 5 6 RRD-H 19 18 17 16 15

www.pearsonhighered.com

PEARSON

Table of Contents

Mini-Mod 1 Whole Numbers 1
Mini-Mod 2 Factors and Fractions 23
Mini-Mod 3 LCM and Fractions 39
Mini-Mod 4 Mixed Numbers 55
Mini-Mod 5 Operations with Decimals 71
Mini-Mod 6 Ratios, Rates, and Percents 89
Mini-Mod 7 U.S. and Metric Measurement 107
Mini-Mod 8 Introduction to Geometry 119
Mini-Mod 9 More Geometry 137
Mini-Mod 10 Statistics 153
Mini-Mod 11 Signed Numbers 173
Mini-Mod 12 Introduction to Algebra 187

Name: ______________________________ Date: ________________

Instructor: ____________________________ Section: ______________

Notebook 1.1
Whole Numbers

OTHER NOTES

____________________ are the numbers
0, 1, 2, 3, 4, 5, 6, 7, 8, 9, 10, 11, 12, 13, 14, …

Note: How many digits are there? ______
List them:

What is the difference between digits and whole numbers?

What system do we use to write whole numbers? ________ __________.

What does that mean? The ________ of each digit in a whole number depends on the ________ as well as its __________ within the number.

Complete the following place value chart:

Each set of three place values forms a ___________.

What is the difference in a digit's place value and its actual value?

EXAMPLE 1 Consider the number 13, 579.
What digit is in the tens place value?

What is the place value of the 3?

What is the actual value of the 3?

We can use __________________________ to better understand the value of each digit in a number.

Expanded notation shows the __________ __________ of each digit based on its __________ __________.

EXAMPLES 2 & 3 Write the following numbers in expanded notation.
4295

245,103

What are the steps for converting numbers to words?

Start at the _____.
Write the number in each period , followed by __________
Do this for each period, …
Do not include the word __________

Note: True/False You must write the comma when writing a 4-digit number.

EXAMPLES 4 & 5 Convert the following numbers to words.
1209

821,504

A number written using digits rather than words is said to be in ________________ ____________.

Write the steps for writing numbers in standard form.

Read the number from ______ to ________.
Write the number in each period, …

EXAMPLES 6 & 7 Write the following numbers in standard form.
twenty-four thousand, seven hundred twelve

Nine thousand, three hundred fourteen

CONCEPT CHECK: How many place values are in a period?

GUIDED EXAMPLES:

1. For 2,053,479, what is the place value of the 5?
2. Write in expanded notation. 14,295
3. Convert to words. 35,697
4. Convert to standard form. Three thousand, two hundred seventy-eight

Name: ______________________ Date: ______________

Instructor: ______________________ Section: ______________

Notebook 1.2
Rounding

OTHER NOTES

What do we mean by rounding?

When we want to get an answer that is _______, but not _______________.

Rounding is finding a number _______________ but _________ to work with.

To round a number, what do you need to know?

EXAMPLE 1 Round 412 to the nearest ten.

EXAMPLE 2 Round 617,725 to the nearest thousand

What is the procedure for rounding?

1. _____________ the digit in the place to which you are rounding.
2. Look at the digit to the __________ of the ___________ digit.
3. If that digit is:

   4 or less, ___________________ the underlined digit.
   This is called _________________ ________________.

   5 or more, ________________ the underlined digit by ________.
   This is called _______________ _______.
4. Replace all digits to the right of the digit ____________________.

EXAMPLE 3 Round to the nearest thousand. 506,243.

EXAMPLE 4 Round to the nearest hundred. 101,697.

Note: Sometimes the digit to the left of the underlined digit must also be changed. When might this occur?

EXAMPLE 5 Round to the nearest ten. 1296

EXAMPLE 6 Round to the nearest hundred. 449,985

CONCEPT CHECK: When rounding a number to the nearest place value, and the number to the right of the digit being rounded is 5, do you round up or round down?

GUIDED EXAMPLES:

1. Round 457 to the nearest ten.

2. Round 19,256 to the nearest thousand.

3. Round 34,996 to the nearest ten.

4. Round 732,819 to the nearest ten thousand.

Name: ______________________________ Date: ________________

Instructor: ___________________________ Section: ______________

Notebook 1.3
Adding Whole Numbers; Estimation

OTHER NOTES

__________ occurs when you combine numbers?

What is the answer to an addition problem?

EXAMPLES 1 & 2 Find the sum.

5 + 6

7+2

What does the Commutative Property of Addition state?
Then give an example.

What does the Associative Property of Addition state?
Then give an example.

What does the Addition Property of Zero state?
Then give an example.

EXAMPLES 3 – 6 For each statement, which property tells us the statement is true? The properties are Commutative and Associative Properties, and the Addition Property of Zero.

2 + 4 = 4 + 2 Property: ___________________

(3 + 5) + 7 = 3 + (5 + 7) Property: ___________________

6 + 0 = 6 Property: ___________________

(1 + 7) + 8 = 8 + (1 + 7) Property: ___________________

Explain the process for adding whole numbers with several digits.

*

*

*

EXAMPLES 7 & 8 Find the sum.

126 + 12

349 + 268 + 1345

What does it mean to estimate?

EXAMPLE 9 Estimate by rounding to the nearest hundred. Then compare the result with the exact answer. 7026 + 13,479

Estimate: Actual:

Write down some key words for phrases for addition:

EXAMPLES 10 - 12 Write the following using symbols, then simplify.

Find 6 increased by 5.

Find 7 more than 10.

Find the sum of 2 and 8.

CONCEPT CHECK: Can you name three words or phrases that indicate addition?

GUIDED EXAMPLES:

1. Add. 137 + 29

2. Add. 4853 + 6484

3. Estimate the total amount spent by rounding to nearest ten dollars.
   $526 + $841

4. Write in symbol form & simplify.
   four increased by six

Name: ______________________ Date: ____________

Instructor: ______________________ Section: ____________

Notebook 1.4
Subtracting Whole Numbers

OTHER NOTES

________________ of two numbers is taking away one number or quantity from another.

What is the answer to a subtraction problem?

EXAMPLE 1 Consider the problem 6 – 2. What is the difference?

Write down the two subtraction facts. Give an example.

1. Any number minus itself is ______________.

2. Any number minus zero is ____________________.

Do you know your addition facts? If not, now is the time to review.

True or False: Every subtraction problem has two related addition problems.

EXAMPLE 2 Rewrite 12 – 5 = 7 as two addition problems.

EXAMPLES 3 & 4 Subtract each of the following. Check by adding.

10 – 4

17 – 2

Write down the steps for subtracting whole numbers with several digits.

*

*

*

*

*

Borrowing is sometimes referred to as ____________________.

EXAMPLE 5 Subtract 57 – 38. Check by adding.

EXAMPLES 6 & 7 Subtract. Check by adding.

232 – 141

2010 – 1963

What is meant by estimation?

EXAMPLE 8 Estimate by rounding to the nearest thousand.
Compare the estimate with the exact answer. 11,976 – 1245

Estimate: Actual:

Write down some key words and phrases for subtraction:

CAUTION!! When translating words, be careful to write the numbers in the correct order. When should the order be written in opposite order?

EXAMPLES 9 & 10 Write the following using symbols, then simplify.

Find 6 decreased by 5.

Find 15 less than 51.

CONCEPT CHECK: Can you name three words or phrases that indicate subtraction?

GUIDED EXAMPLES:

1. Find the difference. 78 – 47
2. Subtract. 137 – 29
3. Estimate by rounding to nearest thousand.
   211,836 – 11,784
4. Write in symbols then simplify.
   6 subtracted from 12.

Name: ______________________________ Date: ________________

Instructor: ____________________________ Section: ______________

Notebook 1.5
Basic Problem Solving

OTHER NOTES

What are the four steps for Problem Solving?

1.

2.

3.

4.

Write down some key words for addition:

Write down some key words for subtraction:

EXAMPLES 1 & 2 Translate the following to symbols and simplify.

The sum of 5 and 3

The difference of 6 and 2

The ______________ of a figure is the distance around the figure.
How do you find the perimeter of a figure?

EXAMPLE 3 A rectangular garden measures 9 feet in length and is 5 feet wide. How many feet of fencing are needed to enclose the garden?

Understand the problem

Create a plan

Find the answer

Check the answer

EXAMPLE 4 A local community college requires 64 credit hours for an Associate Degree. Kylie earned 28 credit hours this past year. How many more credit hours does Kylie need in order to get her degree?

Understand the problem

Create a plan

Find the answer

Check the answer

EXAMPLE 5 On Monday, Alex opened a checking account with an initial deposit of $300. She bought groceries for $75, spent $25 on gas, and spend $20 on new clothes. How much money is in her account after these purchases?

Understand the problem

Create a plan

Find the answer

Check the answer

CONCEPT CHECK: What is the first step in problem solving?

GUIDED EXAMPLES:

1. How much money did Kristen make if she earned $100 but spent $20 and $10?

2. How much money does Jake receive in two weeks if he gets a weekly allowance of $10 and an additional $5 for mowing the lawn?

3. How much money do the Reyher's have left to spend on souvenirs if they started with $200 and bought tickets for $128, parking for $12 & food for $38?

Name: ______________________________ Date: ______________

Instructor: ___________________________ Section: ____________

Notebook 1.6
Multiplying Whole Numbers

OTHER NOTES

_______________ is a shortcut for repeated addition.
What are some ways in which we can write multiplication?

EXAMPLE 1 Tina is scheduled to work 3 days, 10 hours each day.

EXAMPLE 2 Pierre buys 4 dozen eggs at the store.

Do you know your basic multiplication facts?
If not, now is the time to review. Please look in the study guide for an empty multiplication table.

What is the answer to a multiplication problem?

The numbers being multiplied are called __________.

EXAMPLE 3 Identify the factors and find the product? 3×4
Factors: Product:

What does each of the following properties state? Give an example.
Commutative Property of Multiplication -

Associative Property of Multiplication -

Multiplication Property of Zero -

Multiplication Property of One -

Distributive Property of Multiplication over Addition (Distributive Property) –

EXAMPLES 4 & 5 Rewrite using the Distributive Property and simplify.
$8(3 + 6)$ $2(9 + 2)$

Note: The Distributive Property can be used to make mental calculations.
EXAMPLE 6 Multiply 5(23).

EXAMPLES 7–11 For each statement, which property tells us the statement is true? Choose from the following:
Associative Property, Commutative Property, Distributive Property, Multiplication Property of Zero, and Multiplication Property of One.

$6 \cdot 25 = 25 \cdot 6$ Property: ___________________

$5(30 + 2) = 5(30) + 5(2)$ Property: ___________________

$0 \cdot 7 = 0$ Property: ___________________

$6 \cdot (4 \cdot 7) = (6 \cdot 4) \cdot 7$ Property: ___________________

$9 \cdot 1 = 9$ Property: ___________________

Write down the steps for multiplying whole numbers by a single-digit number.

1.

2.

3.

Note: Carrying may also be referred to as ______________________.

EXAMPLE 12 Calculate the following product. $3 \cdot 248$

Write down some key words and phrases for multiplication:

EXAMPLES 13 & 14 Write the following using symbols, then simplify.

Find the product of 2 and 17.

Find 721 times 9.

CONCEPT CHECK: Name two words or phrases that indicate multiplication?

GUIDED EXAMPLES:

1. Rewrite using the Distributive Property. Simplify: $3(8 + 2)$
2. Calculate the product: $157 \cdot 2$
3. Calculate the product: $3(243)$
4. Write in symbol form then simplify: the product of 4 and 5

Name: ______________________ Date: ____________

Instructor: ______________________ Section: ____________

Notebook 1.7
Dividing Whole Numbers

OTHER NOTES

______________ is the operation of splitting a quantity or number into equal parts.

EXAMPLE 1 A group of 15 people goes on a whitewater rafting trip. There are 3 boats, so the group of 15 people needs to be divided into 3 groups of equal size. In this case, each boat will have 5 people, meaning that 15 divided by 3 is 5.

Do you know your basic multiplication facts? You will need them in order to divide. If not, now is the time to review.

Four Basic Division Facts: (Give an example of each.)

1. Any non-zero number divided by itself is __________.
2. Any number divided by 1 is ____________________.
3. Zero divided by any non-zero number is ___________.
4. Division by 0 is ______________________________.

What is the answer to a division problem?

The number being divided is the ______________.

The number you are dividing by is called the _______________.

What are 3 different ways a division problem can be written?

Write the 4 different ways division can be written in symbols.

EXAMPLE 2 Divide the following. $36 \div 4$

The number that is left over from a division problem is the ________.

A number divides exactly into another number if ________________.

Write the steps for performing long division:

1. Set up the problem using the ________ ___________ ___________.
2. _________ the ______________ into part of the ____________.
3. ______________ this result by the _____________.
4.
5. _________ _________the next digit from the ____________.
6. Repeat until all digits in the ___________ have been used.

EXAMPLES 3 & 4 Divide the following.

$92 \div 4$ $\quad\quad\quad\quad$ $3\overline{)148}$

Write down some key words and phrases for division:

CAUTION!! When translating for division, be careful to write down the numbers in the correct order.

EXAMPLES 5 & 6 Write the following using symbols, then simplify.

Find the quotient of 21 and 7.

Find 72 divided by 6.

CONCEPT CHECK: In the problem $\frac{6}{2} = 3$, which number is the quotient?

GUIDED EXAMPLES:

1. Divide.
   $56 \div 7$
2. Divide.
   $\frac{0}{7}$
3. Divide. Find the quotient and remainder.
   $68 \div 9$
4. Write in symbol form, then simplify: the quotient of 24 and 3

Name: ______________________________ Date: ________________

Instructor: ____________________________ Section: ______________

Notebook 1.8
More with Multiplying and Dividing

OTHER NOTES

Write down the steps for multiplying whole numbers with several digits.

1. _______ _______ ____________ vertically so that the digits in the same place values line up.

2. Starting at the ones column, ____________ the bottom digit by each digit in the _____ ___________. Moving left, …

3. For each multiplication, start the _________ in the ________ ________ of the bottom digit. Place __________ to the ________ of each product to fill in all place values of that product.

4. If any product is _______ than _____, write down the _____ digit of that product. Carry the _______ digit to the next place value ____________________________ .

5. Add the _________ __________ to get the overall ________.

EXAMPLES 1 & 2 Calculate the following products.

$21 \cdot 17$ $32 \cdot 125$

What are the steps for multiplying numbers ending in zero(s)?

1.

2.

3.

EXAMPLE 3 Multiply. 260(400)

EXAMPLE 4 Estimate by rounding to the nearest ten. 7227(87)

What are the steps for performing long division?
1.

2.

3.

4.

5.

6.

Note: The final remainder should be less than the ___________________.

EXAMPLE 5 Divide the following. $1482 \div 12$

EXAMPLE 6 Divide the following. $4725 \div 225$

CONCEPT CHECK:
How many zeros are in the final product of 55,000 and 300?

GUIDED EXAMPLES:
Calculate the following products.
1. 157(62)

2. $5700 \cdot 300$

3. Estimate by rounding to the nearest ten. $634 \cdot 27$

4. Divide. $304 \div 23$

Name: ______________________________ Date: ________________

Instructor: ___________________________ Section: ______________

Notebook 1.9
Exponents

OTHER NOTES

An ____________ is used as a shortcut for repeated multiplication.

What does 2^3 mean?

Which is the base? _____ Which is the exponent? ______

CAUTION!!

What is the number, or factor, being multiplied by itself?

The __________________ is the number of times the base is used as a factor.

EXAMPLE 1 Identify the base and exponent, then evaluate.

5^2 Base: ____ Exponent: ____ Value:

Note: For an exponent of 2, we say "_____________________".

For an exponent of 3, we say "_____________________".

EXAMPLES 2 & 3 Evaluate (find the values of) the following.

1^4

6^2

EXAMPLES 4 & 5 Write the following using exponents, then evaluate.

(3)(3)(3)

$7 \cdot 7$

Rule: Any number raised to the power of 1 is __________________.

EXAMPLES 6 & 7 Evaluate.

7^1 13^1

Rule: Any non-zero number raised to the 0 power is _______________.

EXAMPLES 8 & 9 Evaluate.

5^0 79^0

EXAMPLES 10 & 11 Evaluate.

8^2 31^0

Note: Small exponents can result in _________________ very quickly.

EXAMPLE 12 Calculate 10^2, 10^3 and 10^6.

10^2 10^3 10^6

This rapid growth, called ________________________________, occurs in science and business.

CONCEPT CHECK: How can 2^5 be written without exponents?

GUIDED EXAMPLES:

1. Identify the base and exponent.
   5^4
2. Evaluate.
   2^5
3. Write using exponents and evaluate.
   $3 \cdot 3 \cdot 3 \cdot 3$
4. Write using exponents and evaluate.
   (12)(12)

Name: ______________________________ Date: ________________

Instructor: ____________________________ Section: ______________

Notebook 1.10
Order of Operations and Whole Numbers

OTHER NOTES

When there is more than one operation involved, use the Order of Operations so that you will know the order in which to perform the operations.

Write down the steps for using the order of operations:

P:

E:

MD:

AS:

What can you help to remember this order?

EXAMPLE 1 Calculate.

$4 + 6 \cdot 3$

EXAMPLES 2 & 3 Simplify by using the order of operations.

$7 + (13 - 9)$

P:

E:

MD:

AS:

$3^2 + 5 \cdot 4$

P:

E:

MD:

AS:

EXAMPLES 4 & 5 Simplify.

$3(4 - 2)^2 + 5$

$24 \div 3 \cdot 2$

P:

E:

MD:

AS:

Note: What are some other symbols that act like parentheses?

When you see a fraction bar, act like there are parentheses around the ______________________ and ______________________.

For example, what does $\frac{2+1}{4+3}$ mean?

EXAMPLE 6 Simplify. $\frac{22+10}{4\cdot 5-4}$

CONCEPT CHECK:

What operation should be performed first in simplifying $9-5+16\div 2^2$?

GUIDED EXAMPLES:

Simplify.

1. $2+8\times 3-1$

2. $12\div(5-2)+7$

3. $5+(8-2)^2+4$

4. $\frac{20+20}{6+4}$

Name: ______________________________ Date: ________________

Instructor: ___________________________ Section: ______________

Notebook 1.11
More Problem Solving

OTHER NOTES

Recall the four steps for Problem Solving.
1.

2.

3.

4.

Recall the key words and phrases for:
Addition:

Subtraction:

Multiplication:

Division:

EXAMPLES 1–3 Translate the following to symbols.
Then simplify using the order of operations.

twice the sum of 7 and 4

the difference of 9 and 3, divided by 3

three times the product of 2 and 4

The _____ of a ________ is the measure of the surface inside a figure.

How do you find the area of a rectangle?

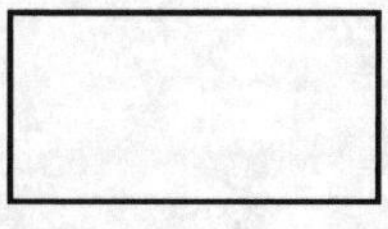

EXAMPLE 4 Find the area of the rectangular room that has a length of 21 feet and a width of 14 feet.
Understand the problem

Create a plan

Find the answer

Check the answer

EXAMPLE 5 Desmond makes a salary of $39,000 per year.
What is his monthly salary?
Understand the problem

Create a plan

Find the answer

Check the answer

EXAMPLE 6 The phone company charges $20 for installation and $25 a month for basic service. How much will Monique pay for one year of service with installation?
Understand the problem

Create a plan

Find the answer

Check the answer

CONCEPT CHECK:
What two steps need to be done before calculation in solving a problem?

GUIDED EXAMPLES:
Translate to symbols, then simplify.

1. 2 more than the product of 3 and 5
2. twice the product of 3 and 7
3. Michelle makes a salary of $54,000 per year. What is her monthly salary?

Name: ______________________________ Date: _______________

Instructor: ___________________________ Section: ______________

Notebook 2.1
Factors

OTHER NOTES

A __________ is a number that is multiplied by another number.

Also, a factor is a whole number that _______ _______ into another number.

What are the factors of 6?

EXAMPLES 1 & 2 List all factors of 8.

List all factors of 24. Hint: Use factor pairs.

A _________ _________ is a number that divides exactly into two or more numbers. Can there be more than one?

What is the largest number that divides exactly into two or more numbers?

What are the steps finding the greatest common factor?

1. List all ____________________.
2. Identify the factors that are __________________.
3. Choose the _________ of these _________ _________.

EXAMPLE 3 Find the greatest common factor, or GCF of 15 and 24.
Factors of 15:

Factors of 24:

Common factors:

Largest:

EXAMPLE 4 Find the greatest common factor, or GCF of 18 and 54.
Factors of 18:

Factors of 54:

Common factors:

Largest:

Rules for Divisibility:

A number is divisible by:	If:	Example:
2	It ends in ...	
3	The sum of ...	
5	It ends in ...	
6	It is divisible by both ...	
9	The sum of ...	
10	It ends in ...	

EXAMPLE 5 Find the GCF of 132 and 198.

Factors of 132:

Factors of 198:

Common factors:

Largest:

What does "to factor" mean?

EXAMPLE 6 List three ways to factor 18.

CONCEPT CHECK: What is a factor?

GUIDED EXAMPLES:

1. Is 1695 divisible by 2? 3? 5? 9? 10?

2. Find the GCF of 12 and 16.

3. Find the GCF of 24 and 56.

4. List as least 2 ways to factor 36.

Name: ______________________________ Date: ________________

Instructor: ___________________________ Section: ______________

Notebook 2.2
Prime Factorization

OTHER NOTES

Factor (noun) – a number that is being ___________.
Factor (verb) – to write as a _____________.

EXAMPLE 1 a. Find all factors of 36. b. Factor 36 several ways.

A _________ _________ is a whole number greater than 1 that has exactly two factors, the number itself and 1.

Is 0 prime? Is 1 prime?

Prime Numbers:

Whole Number	Factors	Prime??
2		
3		
4		
5		
6		
7		
8		
9		

Write down the first 15 prime numbers:

__________ ______________ are whole numbers greater than 1 that are not prime.

Is 0 composite? Is 1 composite?

True/False: Every number is either prime or composite.

EXAMPLE 2 Identify each whole number as prime, composite, or neither.
2, 8, 1, 13, 22, 55, 29, 37, 99

The _________ ____________ of any whole number is the factored form in which all factors are prime numbers.

What are the steps for finding the prime factorization?

1. Write the number as a ______________ of __________.
2. Check to see if each of these factors is _____________.
   If so, ________________; if not, write it as a ________________.
3. Continue to do this until all factors are _________.
4. Write the prime factorization as a product of all the _____________ ________ ___________.

EXAMPLE 3 Find the prime factorization of 24.

EXAMPLE 4 Find the prime factorization of 60.

Note: If factors are listed from smallest to largest, every composite number's prime factorization can be written in ____________________.

CONCEPT CHECK: How many factors does a prime number have?

GUIDED EXAMPLES:

1. Identify as prime, composite, or neither. 55 ________ 17 _________
2. Find the prime factorization of 35.
3. Find the prime factorization of 48.
4. Find the prime factorization of 28.

Name: ______________________________ Date: ________________

Instructor: ____________________________ Section: ______________

Notebook 2.3
Understanding Fractions

OTHER NOTES

What is the difference between whole numbers and fractions?
Give examples of each.

What are numbers that represent the parts of a whole?

EXAMPLE 1 A pizza has 8 slices. Sam ate 3 of the 8 slices for lunch.

The top number of a fraction is called the _____________.

The bottom number of a fraction is called the ________________.

Which one indicates the number of equal parts being considered?

Which one indicates the number of parts in the whole?

EXAMPLE 2 Write a fraction that best represents the shaded part.

a. 

b. 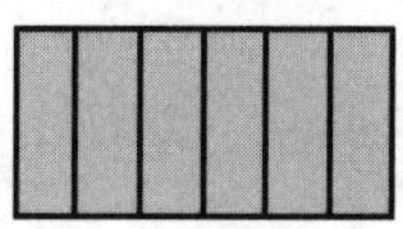

c. 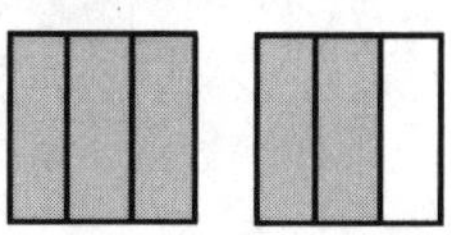

What kind of fraction has the numerator less than the denominator?
Proper or improper??
A proper fraction has a value _______ _______ 1.

Write some examples of proper fractions:

What kind of fraction has the numerator greater than or equal to the denominator? Proper or improper??

An improper fraction has a value ____________________________.

Write some examples of improper fractions:

Note: Any whole number greater than or equal to 1 can be written as an __________ ____________ with the number in the numerator and a ______ in the denominator.

EXAMPLES 3–7 Identify each fraction as proper or improper.

$\frac{4}{9}$

$\frac{11}{4}$

$\frac{7}{6}$

$\frac{3}{3}$

$\frac{0}{5}$

Four Basic Facts about Fractions:

1. Any non-zero number divided by itself is _____.
   Ex.
2. Any number divided by 1 is ________________.
   Ex.
3. Zero divided by any non-zero number is ______.
   Ex.
4. Division by 0 is _________________.
   Ex.

CONCEPT CHECK: Division by zero is _______________.

GUIDED EXAMPLES:

1. Write a fraction that represents the shaded part.

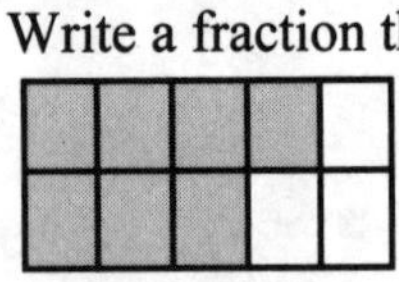

2. Identify $\frac{0}{7}$ as proper or improper
3. Identify $\frac{4}{4}$ as proper or improper.
4. Simplify $\frac{11}{0}$.

Name: ______________________________ Date: _______________

Instructor: ___________________________ Section: _____________

Notebook 2.4
Simplifying Fractions—GCF and Factors Method

OTHER NOTES

Every fraction can be written in many ways.
Write several different ways to represent one-half.

What are equivalent fractions?

A fraction is said to be in ____________ __________
if there is ____ ________ ____________ other than 1
that divides exactly into the ___________ and ______________.

What are the steps to simplify a fraction using the GCF method?

1. Find the _____________ of the numerator and denominator.
2. Divide the ___________ and _____________ by the _____.

True/False: Dividing the numerator and the denominator by the same non-zero number does not change the value of a fraction.

EXAMPLE 1 Write in simplest form using the GCF method. $\frac{9}{15}$

Find the GCF:

Divide num and den by GCF:

EXAMPLE 2 Write in simplest form using the GCF method. $\frac{72}{96}$

Find the GCF:

Divide the num and den by GCF:

If you are having a hard time finding the GCF, use any common factor.

What are the steps for simplifying fractions using the factors method?

1. Find any factor that is _____________________________.
2. Divide the numerator and denominator by ________________.
3. Repeat this until there are ____________________________.

EXAMPLE 3 Write in simplest form using the factors method. $\frac{20}{32}$

EXAMPLE 4 Write in simplest form using the factors method. $\frac{72}{96}$

CONCEPT CHECK: To simplify $\frac{10}{20}$ using the GCF method, what number will be divided into the numerator and denominator?

GUIDED EXAMPLES:

Write in simplest form using the GCF method.

1. $\frac{12}{16}$

2. $\frac{75}{125}$

Write in simplest form using the factors method.

3. $\frac{30}{75}$

4. $\frac{24}{54}$

Name: ______________________________ Date: ______________

Instructor: ___________________________ Section: ______________

Notebook 2.5
Simplifying Fractions—Prime Factors Method

OTHER NOTES

We can use prime factorizations to write a fraction in simplest form.

What are the steps for simplifying fractions using the Prime Factors Method?

1. Write the numerator as a ________________________.
2. Write the denominator as a ______________________.
3. Divide by ____________________________ in the numerator and denominator.
4. Multiply the __________ ___________ to get simplest form.

Note: What results when you divide the numerator and denominator by the same non-zero number?

EXAMPLE 1 Write in simplest form using the prime factors method. $\frac{24}{36}$

Write numerator as a product of primes:

Write denominator as a product of primes:

Divide numerator and denominator by common factors:

Multiply the remaining factors:

EXAMPLE 2 Write in simplest form using the prime factors method. $\frac{84}{91}$

Write numerator as a product of primes:

Write denominator as a product of primes:

Divide numerator and denominator by common factors:

Multiply the remaining factors:

EXAMPLE 3 Write in simplest form using the prime factors method. $\frac{39}{195}$

Write numerator as a product of primes:

Write denominator as a product of primes:

Divide numerator and denominator by common factors:

Multiply the remaining factors:

CONCEPT CHECK: When dividing the numerator and denominator of a fraction by the same non-zero number, what happens to the value of the original fraction?

GUIDED EXAMPLES:

1. Write in simplest form using the prime factors method. $\frac{25}{35}$

2. Write in simplest form using the prime factors method. $\frac{84}{90}$

3. Write in simplest form using the prime factors method. $\frac{42}{126}$

Name: ______________________________ Date: ________________

Instructor: ____________________________ Section: ______________

Notebook 2.6
Multiplying Fractions

OTHER NOTES

How do you find $\frac{1}{2}$ of $\frac{2}{3}$?

Show this.

What are the steps for multiplying fractions?
1.
2.
3.

EXAMPLE 1 Multiply. Write your answer in simplest form. $\frac{4}{7} \cdot \frac{3}{5}$

Multiply numerators.

Multiply denominators.

Simplify.

EXAMPLE 2 Multiply. Write your answer in simplest form. $\frac{7}{10} \cdot \frac{5}{8}$

Multiply numerators.

Multiply denominators.

Simplify.

What are the steps for multiplying fractions by simplifying first?
1.
2.
3.

EXAMPLE 3 Multiply by simplifying first. $\frac{7}{10} \cdot \frac{5}{8}$

Write as one fraction.

Divide by common factors.

Multiply the remaining factors.

EXAMPLE 4 Multiply by simplifying first. $\frac{4}{15}\cdot\frac{9}{16}$

EXAMPLE 5 Multiply by simplifying first. $3\cdot\frac{2}{9}$

EXAMPLE 6 Elizabeth bought a sandwich and ate half of it. She gave her brother one-half of what was left. How much of the sandwich did her brother get?
Understand the problem.

Create a plan.

Find the answer.

Check the answer.

EXAMPLE 7 Multiply by simplifying first. $\frac{2}{8}\cdot\frac{5}{15}$

CONCEPT CHECK: You multiplied fractions to get $\frac{33}{48}$. What will your final answer be?

GUIDED EXAMPLES:

1. Multiply $\frac{2}{3}\cdot\frac{4}{9}$. Write your answer in simplest form.
2. Multiply $\frac{3}{10}\cdot\frac{5}{24}$ by simplifying first.
3. Multiply $\frac{2}{7}\cdot\frac{21}{50}$ by simplifying first.
4. Multiply $\frac{3}{8}\cdot 2$ by simplifying first.

Name: ______________________________ Date: ______________

Instructor: __________________________ Section: ____________

Notebook 2.7
Dividing Fractions

OTHER NOTES

How do you know if two numbers are reciprocals?

How do you find the reciprocal of a fraction?

How do you find the reciprocal of a whole number?

EXAMPLES 1 & 2 Find the reciprocals of the following numbers.

$\frac{3}{4}$

5

What is the reciprocal of 0?

What happens when you divide a number by 2 and multiply the same number by $\frac{1}{2}$?

Rule: Dividing by a fraction is the same as __________________________.

What are the steps for dividing fractions?

1.
2.
3.
4.

EXAMPLE 3 Divide. Write your answer in simplest form. $\frac{3}{5} \div \frac{1}{2}$

Invert the second fraction.

Multiply by the first fraction.

Divide by common factors.

Multiply the remaining factors.

EXAMPLES 4 & 5 Divide. Write your answers in simplest form.

$\frac{4}{5} \div \frac{3}{8}$

$\frac{6}{7} \div \frac{3}{2}$

EXAMPLES 6 & 7 Divide. Write your answers in simplest form.

$12 \div \frac{2}{3}$

$\frac{6}{7} \div 9$

CONCEPT CHECK: How would you rewrite $\frac{3}{8} \div \frac{4}{9}$ as a multiplication problem?

GUIDED EXAMPLES:

Divide each of the following. Write your answer in simplest form.

1. $\frac{2}{3} \div \frac{4}{5}$

2. $\frac{4}{7} \div \frac{8}{11}$

3. $\frac{9}{10} \div 6$

4. $\frac{5}{12} \div \frac{15}{16}$

Name: ______________________ Date: ______________

Instructor: ______________________ Section: ______________

Notebook 2.8
Applications of Multiplying and Dividing Fractions

OTHER NOTES

Write the steps for multiplying fractions by simplifying first.

1.
2.
3.

EXAMPLE 1 Multiply $\frac{3}{5} \cdot \frac{7}{9}$. Write your answer in simplest form.

What are the steps for dividing fractions?

1.
2.
3.
4.

EXAMPLE 2 Divide $\frac{2}{7} \div \frac{4}{5}$. Write your answer in simplest form.

EXAMPLE 3 Sam received $1200 through an academic scholarship. He used two-thirds of it for tuition. How much of the scholarship money was used for tuition?

Understand the problem

Create a plan

Find the answer

Check the answer

EXAMPLE 4 Allen has five-sixths of a pound of small pebbles that are used to cover the bottom of fish bowls. How many fish bowls can he prepare if each bowl uses one-twelfth of a pound of the pebbles?

Understand the problem

Create a plan

Find the answer

Check the answer

EXAMPLE 5 One-fourth of the members of the local chamber choir sing soprano. Of these sopranos, four-fifths of them are female. What fraction of the choir members are female and sing soprano?
Understand the problem

Create a plan

Find the answer

Check the answer

EXAMPLE 6 Megan has three-fourths of a gallon of tea and needs to fill as many glasses as possible. If each glass holds two-thirds of a cup, how many glasses can she fill? Hint: one gallon is the same as 16 cups.
Understand the problem

Create a plan

Find the answer

Check the answer

CONCEPT CHECK: What operation is required to solve this problem? A large recipe that makes 24 servings calls for $\frac{2}{3}$ cups of oil. Andrea plans to make $\frac{1}{4}$ of the recipe.

GUIDED EXAMPLES:

1. Ellie needs to earn 66 credit hours at Roane Mountain College in order to get an Associates degree. She will earn $\frac{1}{6}$ of them this semester. How many credits will she earn this semester?

2. A dump truck holds three-fourths of a ton of mulch. How many flowerbeds can be covered if each flowerbed needs one-eighth of a ton of mulch?

3. Two-thirds of Kylie's allowance is saved for a down payment on a new car. Her parents will match one-fourth of this saved amount. What fraction of Kylie's allowance will her parents match?

4. A cookie recipe calls for three-fourths of a cup of peanut butter. How much peanut butter is needed to make one-half of the recipe?

Name: ______________________________ Date: ________________

Instructor: ____________________________ Section: ______________

Notebook 3.1
Finding the LCM—List Method

OTHER NOTES

EXAMPLE 1 Find the first five multiples of 9.

EXAMPLE 2 Find the first four multiples of 20.

A __________ ____________ of two or more numbers is a number that is a multiple of the given numbers.

True/False You can find a common multiply by multiplying the given numbers together.

The __________ ________________ __________ of two or more numbers is the _______________ number that is a multiple of the given numbers.

What are the steps for finding the LCM of two or more numbers using the List Method?

1.
2.
3.

EXAMPLE 3 Find the least common multiple of 9 and 12.
List the first several multiples of 9.

List the first several multiples of 12.

Identify common multiples.

Choose the least.

EXAMPLE 4 Find the least common multiple of 6, 8, and 12.

List the first several multiples of 6.

List the first several multiples of 8.

List the first several multiples of 12.

Identify common multiples.

Choose the least.

CONCEPT CHECK: What are multiples of 8?

GUIDED EXAMPLES:

1. List the first five multiples of 16.

2. Find the least common multiple of 4 and 6.

3. Find the least common multiple of 12 and 15.

4. Find the least common multiple of 6, 9, and 18.

Name: ______________________ Date: ______________

Instructor: ______________________ Section: ______________

Notebook 3.2
Finding the LCM—GCF Method

OTHER NOTES

What are the steps for finding the LCM of two numbers using the GCF Method?

1.
2.
3.

EXAMPLE 1 Find the LCM of 6 and 9 using the GCF Method.
Find the GCF of 6 and 9.
Multiply 6 and 9.
Divide this product by the GCF.

EXAMPLE 2 Find the LCM of 6 and 15 using the GCF Method.
Find the GCF of 6 and 15.
Multiply 6 and 15.
Divide this product by the GCF.

EXAMPLE 3 Find the LCM of 4 and 18 using the GCF Method.

EXAMPLE 4 Find the LCM of 7 and 9 using the GCF Method.

Note: If the GCF is 1, then the LCM is the

________________________________.

CONCEPT CHECK: Explain how to find the LCM of 10 and 25.

GUIDED EXAMPLES:

1. Find the LCM of 6 and 10 using the GCF method.

2. Find the LCM of 12 and 16 using the GCF method.

3. Find the LCM of 8 and 15 using the GCF method.

Name: ______________________________ Date: ________________

Instructor: ____________________________ Section: ______________

Notebook 3.3
Finding the LCM—Prime Factor Method

OTHER NOTES

What is prime factorization of any whole number?

EXAMPLES 1 & 2

Find the prime factorization of 12.

Find the prime factorization of 42.

What are the steps for finding the LCM of two numbers using the prime factor method?

1.
2.
3.
4.

EXAMPLE 3 Find the LCM of 12 and 42 using the Prime Factor Method.

Prime factorization of 12:

Prime factorization of 42:

LCM:

EXAMPLE 4 Find the LCM of 8 and 30.

Prime factorization of 8:

Prime factorization of 30:

Write the prime factorizations putting the common factors below each other:

Write down the prime factor from each column:

Multiply the lists of primes to get LCM:

EXAMPLE 5 Find the LCM of 18 and 24.

EXAMPLE 6 Find the LCM of 13 and 19.

CONCEPT CHECK: Write the prime factorization of 24.

GUIDED EXAMPLES:

Find the LCM of each pair of numbers using the prime factor method.

1. 15 and 21

2. 20 and 45

3. 72 and 84

Name: ______________________________ Date: ________________

Instructor: ____________________________ Section: ______________

Notebook 3.4
Writing Fractions with an LCD

OTHER NOTES

Things to consider when finding the LCM:
1. If one number divides exactly into the other, then the LCM is …

Give an example:

2. If the numbers have no common factor other than 1, then the LCM is …

Give an example:

3. If the two numbers have a common factor other than 1, then …

Give an example:

What are some methods that can be used to find the LCM?

What is the difference between LCM and LCD?

What is the least common denominator (LCD)?

EXAMPLES 1 & 2

Find the LCM of 4 and 12.

Find the LCD of $\frac{3}{4}$ and $\frac{5}{12}$.

True/False The same process is used to find both the LCD and the LCM.

Fractions are ____________ ________________ if they represent the same value.

Name some other fractions that represent $\frac{1}{6}$.

What are the steps for writing an equivalent fraction?

EXAMPLE 3 Write a fraction equivalent to $\frac{1}{6}$ using a denominator of 12.

List the steps for writing fractions with an LCD.

1.

2.

3.

EXAMPLE 4 Rewrite $\frac{1}{6}$ and $\frac{7}{9}$ using the LCD as the denominator.

EXAMPLE 5 Rewrite $\frac{4}{15}$ and $\frac{5}{6}$ using the LCD as the denominator.

CONCEPT CHECK:

After finding the LCD of $\frac{1}{6}$ and $\frac{2}{7}$, what number will we multiply 6 by to get an equivalent fraction?

GUIDED EXAMPLES:

1. Rewrite $\frac{5}{8}$ using 24 as the denominator.
2. Rewrite $\frac{11}{12}$ and $\frac{7}{18}$ using the LCD as the denominator.
3. Rewrite $\frac{2}{3}$ and $\frac{3}{7}$ using the LCD as the denominator.

Name: ______________________ Date: ______________

Instructor: ______________________ Section: ______________

Notebook 3.5
Adding and Subtracting Like Fractions

OTHER NOTES

Fractions with the same, or common, denominator are called

____________ ________________.

What are unlike fractions?

EXAMPLES 1 & 2

$\frac{1}{8}$, $\frac{3}{8}$, and $\frac{7}{8}$ are ____________ fractions.

$\frac{3}{7}$ and $\frac{3}{5}$ are ____________ fractions.

Write the steps for adding like fractions:

1.
2.
3.

EXAMPLES 3 & 4 Add the following like fractions. Simplify if possible. .

$\frac{2}{5}+\frac{1}{5}$

$\frac{3}{10}+\frac{3}{10}$

Write the steps for subtracting like fractions:

1.
2.
3.

EXAMPLES 5 & 6 Subtract the following like fractions. Simplify if possible.

$\frac{6}{7}-\frac{2}{7}$

$\frac{11}{14}-\frac{5}{14}$

EXAMPLES 7 & 8 Add or subtract the following like fractions. Simplify if possible.

$$\frac{4}{11}+\frac{3}{11}+\frac{5}{11}$$

$$\frac{7}{8}-\frac{3}{8}$$

EXAMPLE 9 Shauna ate $\frac{1}{8}$ of a pizza and Kenzi ate $\frac{3}{8}$ of the same pizza. Together how much of the pizza did the two girls eat?
Understand the problem.

Create a plan.

Find the answer.

Check the answer.

CONCEPT CHECK: Danette walked to the store. First she walked down a street that is $\frac{2}{10}$ of a mile long, and then down another street that is $\frac{5}{10}$ of a mile long. Create a plan to show how far she walked to the store.

GUIDED EXAMPLES:

1. Subtract the like fractions $\frac{7}{10}-\frac{4}{10}$. Simplify if possible.

2. Add the like fractions $\frac{2}{15}+\frac{8}{15}$. Simplify if possible.

3. Subtract the like fractions $\frac{7}{9}-\frac{4}{9}$. Simplify if possible.

4. Add the like fractions $\frac{7}{12}+\frac{7}{12}+\frac{1}{12}$. Simplify if possible.

Name: ______________________________ Date: ________________

Instructor: ____________________________ Section: ______________

Notebook 3.6
Adding and Subtracting Unlike Fractions

OTHER NOTES

What is the first thing you must do to add or subtract unlike fractions?

Caution!! You must have a __________ ______________
when ______________ or ______________ fractions.

Write down the steps for adding and subtracting unlike fractions:

1.

2.

3.

4.

5.

EXAMPLE 1 Add $\frac{3}{4}+\frac{1}{6}$. Simplify if possible.

Find the LCD.

Rewrite each fraction with
the LCD as the denominator.

Add the numerators.
Keep the denominator.
Simplify if possible.

EXAMPLE 2 Add $\frac{3}{8}+\frac{2}{5}$. Simplify if possible.

EXAMPLE 3 Subtract $\frac{5}{6}-\frac{1}{3}$. Simplify if possible.

EXAMPLE 4 Subtract $\frac{7}{10} - \frac{4}{15}$. Simplify if possible.

EXAMPLE 5 Paul ate $\frac{3}{8}$ of an apple pie. His brother Jeff ate $\frac{1}{4}$ of the same pie.

How much more pie did Paul eat than Jeff?

Understand the problem

Create a plan

Find the answer

Check the answer

CONCEPT CHECK:

For the problem $\frac{5}{9} - \frac{1}{3}$, what is the first step?

GUIDED EXAMPLES:

1. Add $\frac{3}{7} + \frac{2}{5}$. Simplify if possible.

2. Add $\frac{1}{14} + \frac{3}{7}$. Simplify if possible.

3. Subtract $\frac{4}{9} - \frac{1}{12}$. Simplify if possible.

4. Subtract $\frac{14}{15} - \frac{3}{5}$. Simplify if possible.

Name: ______________________________ Date: ________________

Instructor: ___________________________ Section: ______________

Notebook 3.7
Order of Operations and Fractions

OTHER NOTES

Recall: When there is more than one operation involved, use the Order of Operations so that you will know the order in which to perform the operations.

Write down the steps for using the order of operations:

P:

E:

MD:

AS:

Remember, certain symbols can act like parentheses. What are they?

EXAMPLE 1 Simplify by using the order of operations. $\left(\frac{1}{3}-\frac{1}{6}\right)+\frac{1}{2}$

P:

E:

MD:

AS:

Simplify:

EXAMPLE 2 Simplify. $\left(\frac{2}{5}+\frac{1}{5}\right)^2 \div \frac{3}{10}$

P:

E:

MD:

AS:

Simplify:

EXAMPLE 3 Simplify. $\frac{1}{4} \div \frac{3}{2} + \frac{1}{2} \cdot \frac{1}{3}$

P:

E:

MD:

AS:

Simplify:

EXAMPLE 4 Simplify by using the order of operations.

$$\left(\frac{3}{4}\right)\left(\frac{1}{2} - \frac{1}{4}\right)^2 + \frac{2}{5} \cdot \frac{1}{2}$$

P:

E:

MD:

AS:

Simplify:

CONCEPT CHECK:

Given the problem, $\frac{1}{2}\left(\frac{3}{4} - \frac{1}{8}\right) + \left(\frac{1}{5}\right)^2$ which operation is performed first?

GUIDED EXAMPLES:

Simplify each of the following.

1. $\frac{3}{5} + \frac{1}{4} \cdot \frac{2}{3} - \frac{1}{10}$

2. $\frac{2}{3} \div \left(\frac{5}{8} + \frac{1}{2}\right) - \frac{1}{9}$

3. $\frac{5}{6} \cdot \frac{1}{2} + \left(\frac{2}{3} \div \frac{4}{9}\right)^2$

Name: ______________________________ Date: ________________

Instructor: ___________________________ Section: ______________

Notebook 3.8
More Applications of Adding and Subtracting Fractions

OTHER NOTES

Write the steps for adding and subtracting unlike fractions.

1.
2.
3.
4.
5.

EXAMPLE 1 Add $\frac{3}{5}+\frac{2}{9}$. Write your answer in simplest form.

EXAMPLE 2 Gabriel spends $\frac{1}{3}$ of his monthly paycheck on rent and $\frac{3}{8}$ of his paycheck on utilities (electric, water, internet, and cell phone). What fraction of his paycheck is spend on both rent and utilities?

Understand the problem

Create a plan

Find the answer

Check the answer

EXAMPLE 3 Helena is selling flowers for homecoming. The florist providing the flowers is to receive one-fourth of the total sales. Helena will receive three-fifths of total sales as her pay, but she needs to pay the florist from her share. How much of her original pay will Helena get to keep?

Understand the problem

Create a plan

Find the answer

Check the answer

EXAMPLE 4 In a class of 18 students, $\frac{1}{3}$ of them earned an A, $\frac{1}{2}$ of them earned a B, and $\frac{1}{6}$ of them earned a C. What fraction of the students earned an A or a B?

Understand the problem

Create a plan

Find the answer

Check the answer

EXAMPLE 5 Bradley Parks is developing a wildflower park that covers three-eighths of an acre. Enough wildflower seed has been purchased to see one-fourth of an acre. How much more of the park still needs to be seeded?

Understand the problem

Create a plan

Find the answer

Check the answer

CONCEPT CHECK: What is the first step in solving the following application? Last weekend it rained $\frac{1}{5}$ of an inch on Saturday and $\frac{7}{9}$ of an inch on Sunday. What was the total rainfall for the weekend?

GUIDED EXAMPLES:

1. Susie is making dinner and needs $\frac{5}{8}$ of a cup of shredded cheese for the casserole and $\frac{1}{3}$ of a cup of shredded cheese for the salad. How much shredded cheese will she need in all?

2. Jeremiah has three-fourths of a gallon of milk and Maggie has two-thirds of a gallon of milk. How much more milk does Jeremiah have than Maggie?

3. On Monday, we received $\frac{3}{8}$ of an inch of rain. On Tuesday, it rained an additional $\frac{1}{4}$ of an inch. What was the total rainfall on Monday and Tuesday?

Name: ______________________________ Date: ______________

Instructor: ___________________________ Section: ____________

Notebook 4.1
Changing a Mixed Number to an Improper Fraction

OTHER NOTES

What is a mixed number?

An improper fraction has a _________________ that is _______________ __________ or ______________ the denominator. It represents at least ___________________.

Write the following as a mixed number and an improper fraction.

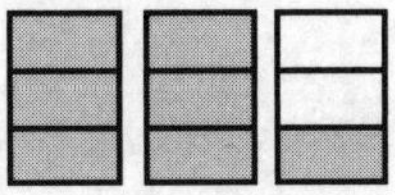

Write down the steps for changing mixed numbers to improper fractions.

1.

2.

3.

EXAMPLE 1 Write $2\frac{1}{3}$ as an improper fraction.

EXAMPLES 2 & 3 Write each as an improper fraction.

$4\frac{5}{7}$ $6\frac{9}{10}$

Write the following picture as a whole number and an improper fraction.

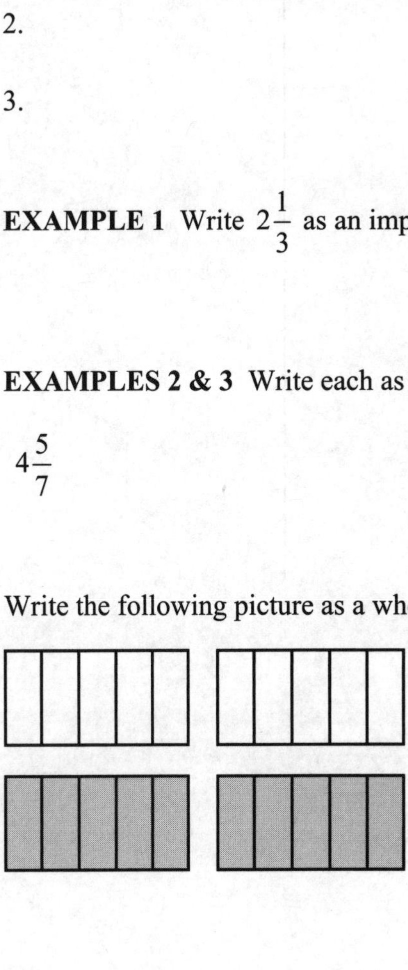

Write down the steps for changing whole numbers into improper fractions.

EXAMPLE 4 Write as an improper fraction. 25

EXAMPLE 5–7 Write each as an improper fraction.

$20\frac{3}{10}$ 91 $2\frac{37}{50}$

CONCEPT CHECK: What is the process for converting 22 to an improper fraction?

GUIDED EXAMPLES:
Write as an improper fraction.

1. $5\frac{3}{8}$

2. $6\frac{2}{3}$

3. $18\frac{4}{5}$

4. 219

Name: ______________________________ Date: ______________

Instructor: ____________________________ Section: ______________

Notebook 4.2
Changing an Improper Fraction to a Mixed Number

Fractions can be represented by ______________________.

EXAMPLE 1 Write $\frac{13}{5}$ as a division problem.

Represent this visually:

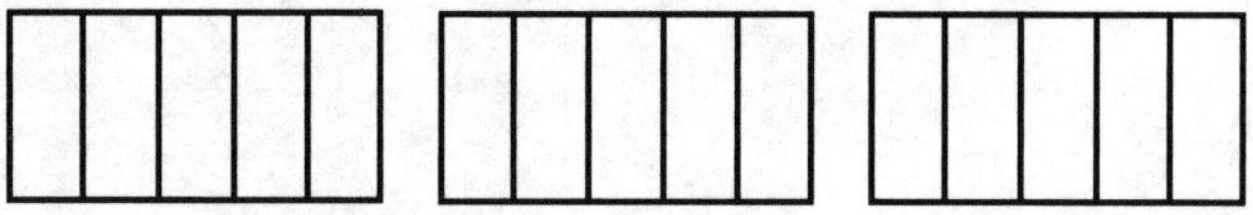

Write the steps for changing improper fractions to mixed numbers.

1.
2. Quotient –

   Remainder –

   Denominator -

3.

CAUTION!! Which part of the fraction goes on the outside of the division bar?

Which part of the fraction goes on the inside?

EXAMPLES 2 & 3 Write each as a mixed number.

$\frac{13}{4}$

$\frac{7}{5}$

Note: What happens if the remainder is 0 when changing improper fractions to mixed numbers?

OTHER NOTES

EXAMPLES 4 & 5 Write each as a mixed or whole number.

$\frac{19}{5}$

$\frac{28}{7}$

CONCEPT CHECK:

Every fraction can be written as a division problem. How do you write $\frac{33}{4}$ as a division problem?

GUIDED EXAMPLES:

1. Write $\frac{35}{4}$ as a mixed number.

2. Write $\frac{60}{8}$ as a mixed number.

3. Write $\frac{56}{7}$ as a mixed number.

4. Write $\frac{99}{11}$ as a mixed number.

Name: ______________________________ Date: ______________

Instructor: ____________________________ Section: ____________

Notebook 4.3
Multiplying Mixed Numbers

OTHER NOTES

Write the steps for multiplying mixed numbers:

1.

2.

3.

EXAMPLE 1 Multiply.

$1\frac{1}{2} \cdot 5\frac{1}{6}$

EXAMPLE 2 Multiply.

$1\frac{2}{3} \cdot 4\frac{1}{5}$

Write as improper fractions:

Multiply the fractions:

Rewrite as a mixed or whole number:

EXAMPLE 3 Multiply.

$3\frac{1}{4} \cdot 2\frac{3}{5}$

Write as improper fractions:

Multiply the fractions:

Rewrite as a mixed or whole number:

EXAMPLE 4 Multiply.

$2\left(3\frac{1}{2}\right)\left(1\frac{2}{7}\right)$

Write as improper fractions:

Multiply the fractions:

Rewrite as a mixed or whole number:

EXAMPLE 5 Ethan ran $2\frac{1}{4}$ miles each day for 3 days. How far did Ethan run during the 3-day period?

Understand the problem

Create a plan

Find the answer

Check the answer

CONCEPT CHECK:

What is the first step for multiplying $1\frac{1}{2}$ and $1\frac{3}{4}$?

GUIDED EXAMPLES:

1. Multiply. $4\frac{2}{5}\cdot 3\frac{1}{2}$

2. Multiply. $5\frac{1}{4}\cdot 1\frac{3}{5}$

3. Multiply. $9\cdot 4\frac{2}{3}$

4. Multiply. $2\frac{3}{8}\cdot 3\frac{2}{3}$

Name: ______________________________ Date: ________________

Instructor: ___________________________ Section: ______________

Notebook 4.4
Dividing Mixed Numbers

Write the steps for dividing mixed numbers:

1.

2.

3.

EXAMPLE 1 Divide. $1\frac{3}{4} \div 2\frac{4}{5}$

EXAMPLE 2 Divide. $3\frac{2}{5} \div 2\frac{2}{3}$

Write as improper fractions:

Invert the second fraction and multiply it by the first:

If improper, rewrite as a mixed or whole number:

EXAMPLE 3 Divide. $4 \div 3\frac{1}{3}$

Write as improper fractions:

Invert the second fraction and multiply it by the first:

If improper, rewrite as a mixed or whole number:

EXAMPLE 4 Divide. $3\frac{2}{5} \div \frac{1}{5}$

Write as improper fractions:

Invert the second fraction and multiply it by the first:

If improper, rewrite as a mixed or whole number:

OTHER NOTES

EXAMPLE 5 Megan purchased $12\frac{1}{4}$ yards of fabric. How many dresses can she make if it takes $2\frac{1}{3}$ yards to make each dress?

Understand the problem

Create a plan

Find the answer

Check the answer

CONCEPT CHECK:

Given the problem $\frac{9}{1} \div \frac{3}{4}$, what is the answer in simplest form?

GUIDED EXAMPLES:

1. Divide. $3\frac{5}{8} \div 5\frac{3}{4}$

2. Divide. $5\frac{1}{3} \div 2\frac{3}{8}$

3. Divide. $12\frac{1}{2} \div 5$

4. Marcus has $7\frac{1}{2}$ cups of flour. His cake recipe calls for $2\frac{1}{2}$ cups of flour. How many cakes can he bake?

Name: ______________________________ Date: ________________

Instructor: ____________________________ Section: ______________

Notebook 4.5
Adding Mixed Numbers

What must be true in order to add fractions?

What is the rule for adding like fractions?

EXAMPLE 1 Add. $3\frac{1}{3}+2\frac{1}{3}$

Write the steps for adding mixed numbers:

1.

2.

3.

4.

EXAMPLE 2 Add. $7\frac{1}{3}+4\frac{3}{5}$

Find the LCD:

EXAMPLES 3 & 4 Add.

$2\frac{5}{6}+9\frac{2}{3}$

$2\frac{1}{3}+15\frac{2}{3}$

EXAMPLES 5 & 6 Add.

$2+9\frac{2}{3}$

$2\frac{1}{6}+\frac{3}{4}$

OTHER NOTES

EXAMPLE 5 Jordan bought $4\frac{2}{3}$ pounds of ham and $1\frac{1}{2}$ pounds of roast beef. How many pounds of lunch meat did Jordan buy?
Understand the problem

Create a plan

Find the answer

Check the answer

CONCEPT CHECK:
What is the first step in adding mixed numbers?

GUIDED EXAMPLES:

1. Add. $5\frac{4}{9}+3\frac{8}{9}$

2. Add. $35\frac{1}{2}+\frac{3}{10}$

3. Add. $24\frac{1}{2}+103\frac{1}{2}$

4. Aleta made a quilt using $1\frac{3}{4}$ yards of blue fabric and $3\frac{1}{2}$ yards of green fabric. How many yards of fabric did she use for the quilt?

Name: ______________________________ Date: ________________

Instructor: ____________________________ Section: ______________

Notebook 4.6
Subtracting Mixed Numbers

OTHER NOTES

What must be true in order to subtract fractions?

What is the rule for subtracting like fractions?

EXAMPLE 1 Subtract. $5\frac{4}{7}-2\frac{1}{7}$

Write the steps for subtracting mixed numbers:
1.

2.

3.

4.

Notes: When is borrowing needed?

What is another name for "borrowing"?

EXAMPLE 2 Subtract. $6\frac{1}{2}-3\frac{1}{4}$
Is borrowing necessary?

EXAMPLE 3 Subtract. $9\frac{1}{3}-2\frac{5}{8}$
Is borrowing necessary?

EXAMPLES 4 & 5 Subtract.

$10\frac{2}{5}-3$

$4-1\frac{1}{2}$

EXAMPLE 5 James needs a piece of wood $4\frac{1}{3}$ feet long and buys a 8-foot board. If he cuts a piece $4\frac{1}{3}$ feet long, how long is the leftover piece of wood?

Understand the problem

Create a plan

Find the answer

Check the answer

CONCEPT CHECK: Consider the problem $4\frac{1}{6}-2\frac{5}{6}$. What fraction should be added to the fraction $\frac{1}{6}$?

GUIDED EXAMPLES:

Subtract the following.

1. $4\frac{5}{9}-2\frac{2}{9}$

2. $5\frac{1}{2}-2\frac{5}{6}$

3. $17\frac{3}{5}-8$

4. $7-2\frac{2}{9}$

Name: ______________________ Date: ______________

Instructor: ______________________ Section: ______________

Notebook 4.7

Adding and Subtracting Mixed Numbers—Improper Fractions

OTHER NOTES

This is an alternative method for adding and subtracting mixed numbers. This method involves rewriting all mixed numbers as

______________________.

EXAMPLE 1 Add. $2\frac{4}{5}+1\frac{2}{5}$

EXAMPLE 2 Subtract. $3\frac{4}{5}-2\frac{2}{5}$

Write the steps for adding and subtracting mixed numbers using improper fractions:

1.

2.

3.

4.

5.

EXAMPLE 3 Add. $4\frac{2}{3}+3\frac{2}{3}$

EXAMPLE 4 Subtract. $5\frac{1}{8}-3\frac{3}{8}$

EXAMPLE 5 Add. $1\frac{2}{5}+3\frac{1}{2}$

EXAMPLE 6 Subtract. $5\frac{1}{3}-3\frac{5}{6}$

CONCEPT CHECK:

In the problem $6\frac{5}{6}-4\frac{1}{8}$, what is the LCD?

GUIDED EXAMPLES:

1. Subtract. $8\frac{2}{5}-3\frac{4}{5}$

2. Add. $4\frac{1}{9}+2\frac{5}{9}$

3. Add. $4\frac{1}{3}+3\frac{1}{4}$

4. Subtract. $5\frac{1}{2}-1\frac{1}{6}$

Name: ______________________________ Date: ________________

Instructor: ___________________________ Section: _______________

Notebook 4.8

Applications of Operations with Mixed Numbers

OTHER NOTES

Write the steps for multiplying mixed numbers

1.
2.
3.

Write the steps for dividing mixed numbers.

1.
2.
3.

EXAMPLE 1 Divide $6\frac{2}{3} \div 1\frac{3}{5}$. Write your answer in simplest form.

Write the steps for adding mixed numbers.

1.
2.
3.
4.

Write the steps for subtracting mixed numbers.

1.
2.
3.
4.

EXAMPLE 2 Subtract $9\frac{2}{5} - 6\frac{3}{7}$. Write your answer in simplest form.

EXAMPLE 3 Kolton pitched $4\frac{2}{3}$ innings in game 1 of the district tournament.

He then pitched $6\frac{2}{3}$ innings in the next game. How many total innings did Kolton pitch in these two games?

Understand the problem

Create a plan

Find the answer

Check the answer

EXAMPLE 4 It takes Max approximately $4\frac{1}{2}$ days to paint the inside of an average-sized house. About how many such houses can he paint during a 23-day pay period?

Understand the problem
Create a plan
Find the answer
Check the answer

EXAMPLE 5 Kimber can crochet $1\frac{1}{3}$ scarves from one spool of yarn. How many spools of yarn will she need in order to crochet $8\frac{1}{2}$ scarves?

Understand the problem
Create a plan
Find the answer
Check the answer

EXAMPLE 6 On Thanksgiving Day, the 5K Turkey Trot is run. The length of this race works out to around $3\frac{1}{10}$ miles. Marty runs the first $2\frac{1}{2}$ miles in 24 minutes. How much farther does she have to run to finish the race?

Understand the problem
Create a plan
Find the answer
Check the answer

CONCEPT CHECK: Heath is jogging $4\frac{3}{5}$ of a mile and has $1\frac{2}{3}$ miles left. How far has he run? Will this calculation require finding an LCD. Why or why not?

GUIDED EXAMPLES:

1. During registration, the advising center will be open for $7\frac{1}{2}$ hours with one advisor on duty. How many advisors will be needed if each advisor works a $1\frac{1}{2}$-hour shift?

2. The recipe for making one gallon of sweet tea calls $1\frac{1}{3}$ cups of sugar. How much sugar is needed to make $4\frac{1}{2}$ gallons of sweet tea?

3. Raekwon spent $15\frac{3}{4}$ hours volunteering at the local food pantry. If his goal is to volunteer 35 hours this semester, how many more hours does he need to spend?

Name: ______________________ Date: ______________

Instructor: ______________________ Section: ______________

Notebook 5.1
Decimal Notation

OTHER NOTES

What are the three parts of a decimal number?

-

-

-

EXAMPLE 1 Identify the three parts of the decimal number 4.75.

What does the decimal part of a decimal number represent?
Then give some examples.

EXAMPLE 2 For each figure, write the decimal fraction and the decimal number that represents each shaded part.

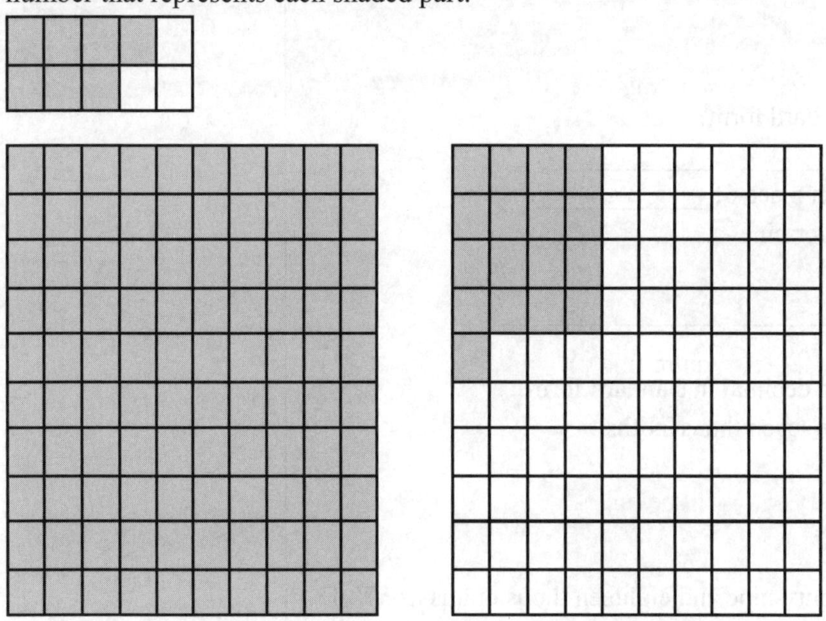

Whole Numbers				.	Decimals				
One Thousands	Hundreds	Tens	Ones	Decimal point	Tenths	Hundredths	One-Thousandths	Ten-Thousandths	Hundred-Thousandths

EXAMPLE 3 Read the decimals.

2.4

214.79

EXAMPLES 4–6 Read the following decimals.

2.3

4.17

12.0457

What are the steps for writing decimal in words?

1. Write the __________ __________ in words.
2. Write the word ________ for the decimal point.
3. Write the __________ ________ of the number in words.
4. Follow this by the __________ ________ of the last __________.

EXAMPLES 7 – 9 Write each decimal in words.

0.32

15.703

2.25

How do you write decimals in standard form?

1. Write the whole number in ________ __________.
2. Write the ______________ in place of __________.
3. Use the given place value to determine
   ____________________________.
4. Write the decimal part in __________ ______________-.

EXAMPLES 11 & 12 Write each decimal in standard form.

twenty-one and two hundred thirty-seven thousandths

fourteen and eight hundredths

CONCEPT CHECK:

Amanda wrote a five hundred seventy-nine and eighteen thousandths as 578.18. This is incorrect. How can she correct it?

GUIDED EXAMPLES:

1. Write the fraction and decimal represented by the picture.

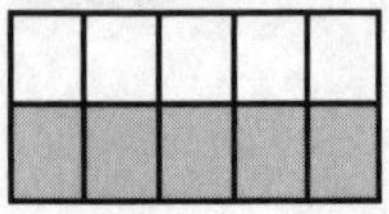

2. Write in words. 4.128

Name: ______________________________ Date: ________________

Instructor: ____________________________ Section: ______________

Notebook 5.2
Comparing Decimals

OTHER NOTES

Which is larger, 0.3 or 0.7?

EXAMPLES 1 & 2 Of each pair of decimals, which is larger?
0.14 and 0.41

2.315 and 2.871

Define "less than".

Define "greater than".

Note: Which number does the symbol point to?

EXAMPLE 3 Which is larger, 0.6 or 0.59?

Rule: What does "adding zeros to the end of a decimal" do to a number?

EXAMPLE 4 0.5 is the same as 0.50.

CAUTION!! Does inserting zeros at the beginning of a decimal part change its value? Give an example.

Write the steps for comparing decimals:
1.
2.
3.

EXAMPLES 5 & 6 Fill in the blank with < or > to make a true statement.
4.13 _____ 4.90

0.983 _____ 0.7822

EXAMPLE 7 Arrange these decimals in order from smallest to largest.
1.23, 1.31, 1.2, 1.0567

EXAMPLE 8 Order the scores from smallest to largest.
During the 2998 Olympic Games, the U. S. Women Gymnastics Team scored 46.875 on the vault, 47.975 on the uneven bars, 47.25 on the balance beam, and 44.425 on the floor exercises.

CONCEPT CHECK:
True/False 2.6 < 5.4

GUIDED EXAMPLES:
1. Fill in the blank with <, > or =. 0.46 _____ 0.45

2. Fill in the blank with <, > or =. 0.251 _____ 0.25

3. Fill in the blank with <, > or =. 0.94 _____ 0.9324

4. Order from smallest to largest. 6.1, 5.23, 5.2

Name: ______________________________ Date: ________________

Instructor: ___________________________ Section: ______________

Notebook 5.3
Rounding Decimals

OTHER NOTES

EXAMPLE 1 Rasheed buys a camera for $328.28. There is a sales tax, which calculates to $20.5175. The store needs to round this tax to the nearest cent, or hundredths place, before adding it to Rasheed's bill.

Whole Numbers				.	Decimals				
One Thousands	Hundreds	Tens	Ones	Decimal Point	Tenths	Hundredths	One-Thousandths	Ten-Thousandths	Hundred-Thousandths

EXAMPLES 2–4 Each of the following often involves rounding of decimals:

*

*

*

Write the steps for rounding decimals to a given place value.

1. Underline the digit in the …

2. Look at the digit to the ________ of the underlined digit.

3. If that digit is:
   - 4 or less, …

   - 5 or more, …

4. Leave off all digits …

EXAMPLE 5 Round 13.45812 to the nearest thousandth.
What digit is underlined?
What is the digit to the right?
Round up or down?
Which digits are left off?

EXAMPLE 6 Round 6.7481 to the nearest hundredth.

Note: When might it be necessary to change the digit to the left of the underlined digit?

EXAMPLE 7 Round 7.9561 to the nearest tenth.

CONCEPT CHECK:

The sales tax for an item is $3.125 and needs to be rounded to the nearest cent. What number needs to be underlined?

GUIDED EXAMPLES:

1. Round to the nearest tenth. 4.65
2. Round to the nearest hundredth. 19.697
3. Round to the nearest thousandth. 3.1282
4. Round to the nearest tenth. 123.571

Name: ______________________________ Date: ______________

Instructor: ___________________________ Section: ____________

Notebook 5.4
Adding and Subtracting Decimals

OTHER NOTES

How is adding and subtracting decimals similar to adding and subtracting whole numbers and like fractions?

EXAMPLE 1 Jennifer spent \$2.97 for a meal and \$1.19 for dessert, including tax. How much did she spend in total?

What is the procedure for adding and subtracting decimals?

1. Write the numbers __________, …
2. Write all decimals with …
3. Add or subtract, starting …
4. Place the decimal point …

EXAMPLES 2 & 3 Add. Check by estimating.

3.27 + 15.2

4 + 12.3 + 0.571

EXAMPLES 4 & 5 Find the difference. Check by estimating.

13.24 – 7

21.5 – 16.43

What is perimeter of a triangle?

How do you find the perimeter of a triangle?

EXAMPLE 5 Find the perimeter of the triangle.

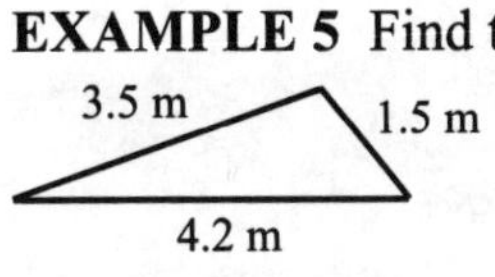

EXAMPLE 6 At Sandee's Malt Shop, a hamburger costs \$3.95, a hot dog costs \$2.49, French fries cost \$1.59, and onion rings cost \$1.89. How much more does a hamburger with French fries cost than a hot dog with onion rings?

Understand the problem

Create a plan

Find the answer

Check your answer

CONCEPT CHECK:

A triangle has sides of length 10.9 in., 8.2 in., and 8.1 in. What is the proper expression for finding its perimeter?

GUIDED EXAMPLES:

1. Add. 20.52 + 6.9 + 13.8

2. Add. 9 + 6.501 + 43.2

3. Subtract. 246.72 – 98.264

4. Subtract. 62.2 – 51.98

Name: ______________________________ Date: ________________

Instructor: ___________________________ Section: ______________

Notebook 5.5
Multiplying Decimals

OTHER NOTES

Show why multiplying $\frac{3}{10}$ and $\frac{2}{10}$ is the same as multiplying 0.2 and 0.3.

Write the steps for multiplying decimals.

1.

2.

3.

EXAMPLES 1 & 2 Multiply.

3.45(6.1)

0.152×0.23

EXAMPLES 3 & 4 Multiply.

3.2×0.008

4 (2.5)

What is the procedure for multiplying a decimal by a power of 10?

What do you do if the number of zeros is more than the number of original decimal places?

EXAMPLE 5 Multiply.

5.793×100

EXAMPLES 6 & 7 Multiply.

79.251 × 10

42.8 × 1000

EXAMPLE 8 Ian's new car travels 25.6 miles for each gallon of gas. How many miles can he travel on a full tank, which contains 12 gallons of gas?

Understand the problem

Create a plan

Find the answer

Check your answer

CONCEPT CHECK:

Nataliya jogs seven days a week along a running path that is 2.9 miles long. Her goal is to jog 20 miles a week. Does she achieve her goal by jogging this path every day?

GUIDED EXAMPLES:

1. Multiply. 1.5 × 3.1

2. Multiply. 0.035 × 0.08

3. Multiply. 68.157 × 100

4. Multiply. 9.63 × 10,000

Name: ______________________________ Date: ________________

Instructor: ___________________________ Section: ______________

Notebook 5.6
Dividing Decimals

OTHER NOTES

What is the number being divided?
What is the number the dividend is being divided by?
What is the answer to a division problem?

Label each part of a division problem in three different ways:

What is the rule for dividing a decimal by a whole number?

EXAMPLE 1 Divide. $6.3 \div 3$

EXAMPLE 2 Simon, Penny, Ashley, and James went out to eat. The total bill was \$36.24. If they split the bill evenly, how much did each person pay?
Understand the problem

Create a plan

Find the answer

Check the answer

EXAMPLE 3 Divide. $6.7 \div 10$

Write the rule for dividing a decimal by a power of 10.

What do you do if the number of zeros is more than the number of whole number places?

EXAMPLE 4 Divide. $2.9 \div 1000$

What happens when we divide 3.8 by 4?
When working with decimals, there are no ____________________.

What are the three possibilities when dividing decimals?
1.

2.

3.

EXAMPLE 5 Divide 6.1 by 3.

Write the steps for dividing a decimal by a decimal.
1.

2.

3.

4.

EXAMPLE 6 Divide. $0.0612 \div 0.12$

EXAMPLE 7 Divide. Round to the nearest hundredth. $1.219 \div 0.3$

CONCEPT CHECK:
In the problem $19.21\overline{)24.567}$, what does the dividend become after the divisor has been changed to a whole number?

GUIDED EXAMPLES:
1. Divide. $8.64 \div 4$

2. Divide. $245.967 \div 1000$

3. Divide. Round to the nearest tenth. $1.4 \div 0.6$

4. Divide. $5.564 \div 4.28$

Name: ______________________________ Date: ________________

Instructor: ____________________________ Section: ______________

Notebook 5.7
Order of Operations and Decimals

OTHER NOTES

Write down the steps for using the order of operations:

P:

E:

MD:

AS:

Remember, certain symbols can act like parentheses. What are they?

What are the two memory tips for remembering the order of operations?

EXAMPLES 1 & 2 Simplify.

$3.6+(0.2+0.3)^2$

P:

E:

MD:

AS:

$(12.6-11.4)^2 \div (0.2)^3$

P:

E:

MD:

AS:

EXAMPLES 3 & 4 Simplify.

$10.5 - 3.2 + 7.2 \div 3.6$

P:

E:

MD:

AS:

$10 \div 5(0.1)^2$

P:

E:

MD:

AS:

Note: A fraction bar can also act like ______________________________.

EXAMPLE 5 Simplify. $\dfrac{10.1 + 2.1(3)}{5.05 + 3.15}$

P:

E:

MD:

AS:

CONCEPT CHECK:

What is the second calculation involved in evaluating $19.12 - 9.6 \div 1.2 \times 1.5$?

GUIDED EXAMPLES:

1. Simplify. $7.1 - (0.6)(0.5 \div 125)^2$

2. Simplify. $12.6 \div 3 + (1.8 + 0.4)^2$

3. Simplify. $2.1 + 8.4 \times 3.12 - 1.37$

Name: ______________________________ Date: ________________

Instructor: ____________________________ Section: ______________

Notebook 5.8
Converting Fractions to Decimals

OTHER NOTES

Every fraction can be written as, or converted to, an _________________________.

Write $\frac{57}{100}$ as an equivalent decimal:

EXAMPLES 1 – 3 Write each fraction as an equivalent decimal.

$\frac{1}{10}$ $\frac{87}{100}$ $\frac{119}{1000}$

Does the denominator have to be a power of 10 in order to write the fraction as a decimal?

What are some common fractions and their decimal equivalents?

$\frac{1}{4}$ $\frac{1}{2}$ $\frac{3}{4}$

How do you write a fraction as a decimal?

EXAMPLES 4 & 5 Write the decimal as a fraction in simplest form.

$\frac{3}{5}$

$\frac{7}{8}$

Caution! The _____________ is being divided by the _____________.

EXAMPLE 6 Convert the fraction to a decimal.

$\frac{2}{3}$

What do we mean by a repeating decimal? What does the bar represent?

EXAMPLE 7 Convert the fraction to a decimal.
Round to the nearest thousandth.

$\frac{4}{11}$

EXAMPLES 8 & 9 Convert the fraction or mixed number to a decimal.

$2\frac{3}{4}$

$\frac{11}{4}$

EXAMPLE 10 The lead-off batter for the Raiders baseball team got on base 31 out of 70 times at bat. Write this fraction as a decimal rounded to the nearest thousandth.

CONCEPT CHECK:

Which procedure can be used to convert $5\frac{3}{4}$ to a decimal?

GUIDED EXAMPLES:

1. Convert the fraction to a decimal. $\frac{4}{5}$

2. Convert the fraction to a decimal. $\frac{1}{4}$

3. Convert the fraction to a decimal. $\frac{5}{8}$

4. Convert the fraction to a decimal. Round to the nearest hundredth. $\frac{2}{7}$

Name: ______________________ Date: ____________

Instructor: ____________________ Section: ____________

Notebook 5.9
Converting Decimals to Fractions

Write 0.9 as a fraction:

Write 0.23 as a fraction:

What are the steps for writing a decimal as a fraction or mixed number?

1.

2.

3.

4.

Note: What do you know about the number of decimal places and the number of zeros in the power of 10 in denominator?

EXAMPLE 1 Write 5.277 as a fraction in simplest form.

EXAMPLES 2 & 3 Write the decimal as a fraction in simplest form.
0.96

0.0201

EXAMPLES 4 & 5 Write the decimal as a fraction in simplest form.
0.425

0.02001

OTHER NOTES

EXAMPLES 6 & 7 Write the decimal as a fraction in simplest form.
4.0358

5.2

CONCEPT CHECK:
How many zeros will be in the denominator of 6.423 when written as a fraction?

GUIDED EXAMPLES:
Write as a fraction in simplest form.

1. 0.5

2. 0.59

3. 0.256

4. 10.1281

Name: ______________________________ Date: ________________

Instructor: ____________________________ Section: ______________

Notebook 6.1
Ratios

OTHER NOTES

What is a ratio?

What are the different ways in which a ratio can be written?
Give examples.

EXAMPLE 1 A gallon of lemonade is made by mixing 2 cups of lemon juice with 14 cups of water. Write the ratio of lemon juice to water in a gallon of lemonade.

What is the rule for writing ratios as fractions?

The quantity appearing first is …

The quantity appearing second is …

EXAMPLE 2 Write 5 hours to 7 hours as a ratio. Use all three ways.

Do we need to include the units in ratios? ______ Why or why not?

EXAMPLES 3 & 4 Write the following ratios as fractions in simplest form.

5 inches to 11 inches

4 ounces: 32 ounces

EXAMPLES 5 & 6 Write the following ratios as fractions in simplest form.

$1.23 to $3.75

Joseph lost $5\frac{1}{2}$ pounds, while his friend Demarcus lost $3\frac{1}{4}$ pounds. Find the ratio of the weight that Joseph lost to the weight that Demarcus lost.

EXAMPLE 7 A cake recipe requires $2\frac{1}{2}$ cups of flour and $1\frac{3}{4}$ cups of sugar. Write the ratio of the amount of flour to the amount of sugar as a fraction in the simplest form.

Understand the problem

Create a plan

Find the answer

Check the answer

CONCEPT CHECK: True/False The quantity appearing second appears in the numerator.

GUIDED EXAMPLES:

Write each ratio as a fraction in simplest form.

1. 11 inches to 9 inches

2. $1\frac{1}{2}$ hours to $3\frac{1}{2}$ hours

3. $3\frac{3}{4}$ gallons to $4\frac{1}{2}$ gallons

4. Mark's truck can tow 2.5 tons and Kyle's truck can tow 5.0 tons. Find the ratio of the towing capacity of Mark's truck to the towing capacity of Kyle's truck.

Name: ______________________________ Date: ______________

Instructor: ___________________________ Section: ____________

Notebook 6.2
Rates

OTHER NOTES

What is a rate?

What are two common examples of rates?

What is the rule for writing rates as fractions?

The quantity appearing first is …

The quantity appearing second is …

Do we need to include the units in rates? ______ Why or why not?

EXAMPLE 1 Write the rate as a fraction in simplest form.

240 miles per 10 gallons

EXAMPLE 2 Write the rate as a fraction in simplest form.

18 tomatoes for 4 pots of stew

EXAMPLES 3 & 4 Write the following ratios as fractions in simplest form.

20 door prizes for 110 people

$250 for 20 hours

What is a unit rate?

What is the difference between a rate and a unit rate?

EXAMPLE 5 Find the unit rate.

A 6-pack of juice bottles costs $3.00. Find the unit rate of cost per 1 bottle.

Note: To find a rate, write the rate as …

To find a unit rate, write the rate as …

What is the procedure for finding the better buy?

EXAMPLE 6 Which of the following is a better buy?
A 12-ounce box of Antonio's Macaroni costs $2.08.

A 15-ounce box of Sally's Macaroni costs $3.05.

Solution: Find the cost per ounce for each box:

What is the cost per ounce for a 12-ounce box?

What is the cost per ounce for a 15-ounce box?

Which one is the better buy?

CONCEPT CHECK: True/False The best buy has the highest unit cost.

GUIDED EXAMPLES:
Write the rate as a fraction in simplest form.

1. 15 departments for 10 salespeople

2. $210 for 28 pizzas

3. Find the unit rate of miles per gallon. 118 miles for 5 gallons

4. Murphy's Crunch cereal comes in a 16-ounce box for $3.20 and Dawn's Nuggets cereal comes in a 25-ounce box for $6.00. Which is the better buy?

Name: ______________________________ Date: ________________

Instructor: ___________________________ Section: ______________

Notebook 6.3
Proportions

OTHER NOTES

What is an equation?

What is a proportion?

EXAMPLES 1 & 2 For each example, write as a proportion.

3 is to 6 as 12 is to 24

Caleb paid \$60 for 4 tickets and Beth paid \$90 for 6 tickets.

CAUTION!! Is consistency important when setting up a proportion?
What does that mean?

What is the procedure for determining if a statement is a proportion?

If the cross products are equal, then the statement is …
If the cross products are not equal, then the statement is …

EXAMPLES 3 & 4 Determine if the statement is a proportion.

$$\frac{6}{10} = \frac{10}{35}$$

$$\frac{2}{3.1} = \frac{3}{3.7}$$

If we know three of the four numbers in a proportion, $\frac{a}{b} = \frac{c}{d}$, such that $b \neq 0$ and $d \neq 0$, can we find the fourth number?
Does it matter which one is missing?

EXAMPLE 5 Find the missing number in the proportion.

$$\frac{3}{9} = \frac{4}{?}$$

Write the steps for finding missing number in a proportion:

1. Calculate the ________________________ and set them ____.

2. Divide both sides of the equation by …

EXAMPLE 6 Find the missing number in the proportion.

$$\frac{5}{6} = \frac{n}{12}$$

Check:

EXAMPLE 7 The proper dosage of a children's medicine is 40mg for every 50 pounds. How much medicine do you give a child weighing 75 pounds?
Understand the problem

Create a plan

Find the answer

Check the answer

CONCEPT CHECK: What symbol does ALL equations contain?

GUIDED EXAMPLES:

1. Write the following as a proportion.
   7 is to 12 as 14 is to 24

2. Determine if the statement is a proportion.
   $$\frac{3.6}{4.8} \stackrel{?}{=} \frac{2.4}{3.2}.$$

3. Find the missing number In the proportion.
   $$\frac{5}{7} = \frac{n}{28}$$

4. You can drive 200 miles on 8.5 gallons of gas. How much gas will it take to drive 600 miles?

Name: ______________________ Date: ______________

Instructor: ______________________ Section: ______________

Notebook 6.4
Percent Notation

What does percent mean?

What are some examples of percents?

How do you write a percent as a fraction?

EXAMPLES 1–3 Write each percent as a fraction in simplest form.

47%

14%

136%

When is it easy to convert fractions to percents?

EXAMPLES 4 & 5 Write each fraction as a percent.

$\frac{7}{10}$

$\frac{61}{100}$

EXAMPLES 6–8 Answer the following questions involving percents.
Penny answered 87 out of 100 problems correctly on her math test. What percent of the problems did she answer correctly?

68 out of 100 cars in a parking lot are white. What percent of the cars are white?

46 out of 100 children prefer strawberry ice cream. What percent of children do NOT prefer strawberry ice cream?

OTHER NOTES

What does 100% represent?

What does more than 100% mean?

What does less than 100% mean?

EXAMPLES 9–12 Are the following common percents less than, equal to, or greater than one whole unit?

50%	225%
100%	150%

What is the procedure for finding the percent of a number using fractions?

To find *p*% of a number, …

Find 4% of 200.

EXAMPLE 13 A company produced 2500 personal computers, and 3% of them were defective. How many of the computers were defective?

Understand the problem

Create a plan

Find the answer

Check the answer

CONCEPT CHECK: Which percent below represents more than one whole unit?
99%, 2%, 101% or 50% ?

GUIDED EXAMPLES:

1. Write 62% as a fraction in simplest form.
2. Write $\frac{83}{100}$ as a percent.
3. Jackie has 100 marbles and 29 of them are red. What percent of the marbles are red?
4. A company produced 5000 digital cameras and 1% of them were defective. How many of the cameras were defective?

Name: ______________________________ Date: ______________

Instructor: ____________________________ Section: ____________

Notebook 6.5
Percent and Decimal Conversions

OTHER NOTES

Recall: What does percent mean?

EXAMPLE 1 Write 51% as a decimal.

What is the procedure for writing a percent as a decimal?

Note: What do you do if there is no decimal point in the original percent?

CAUTION!! It may be necessary to insert zeros in order to …

EXAMPLES 2 – 5 Write each percent as a decimal.

135%

24.9%

$74\frac{1}{2}\%$

4.5%

EXAMPLE 6 Write 0.71 as a percent.

What is the procedure for writing a decimal as a percent?

CAUTION!! It may be necessary to add zeros in order to …

EXAMPLES 7–10 Write each decimal as a percent.

2.3

0.095

0.75

0.004

What is the procedure for finding the percent of a number?

For example, find 10% of 500.

EXAMPLE 11 A company made 7500 tables, and 20% of these tables were made from oak. How many of the tables were made from oak?

Understand the problem

Create a plan

Find the answer

Check the answer

CONCEPT CHECK: Carrie converted 0.53 to a percent and got an answer of 0.0053%. What was her mistake?

GUIDED EXAMPLES:

Write the percent as a decimal.

1. $9\frac{3}{4}\%$

2. $87\frac{1}{4}\%$

Write the decimal as a percent.

3. 0.225

4. A company made 6000 folders, and 5% of the folders were red. How many of the folders were red?

Name: ______________________________ Date: ________________

Instructor: ___________________________ Section: ______________

Notebook 6.6
Percent and Fraction Conversions

OTHER NOTES

Recall: What does $p\%$ mean?

Recall: What are two ways you can convert $p\%$ to a fraction?

1.

2.

Note: Dividing p by _____ and multiplying p by ______ both result in the same answer.

EXAMPLE 1 Write 8% as a fraction in simplest form.

EXAMPLES 2–4 Write each percent as a fraction in simplest form.

Joe sells cars and receives $3\frac{1}{4}\%$ on each sale.

Property tax is $\frac{3}{10}\%$ off the estimated value.

The sales tax is 4.5% of the purchase price.

How do you write a fraction as a percent?
If the denominator is 100:

If the denominator is not 100:

EXAMPLES 5 & 6 Write each fraction as a percent.

$\frac{32}{100}$

$\frac{5}{8}$

EXAMPLE 7 Write as a percent. Round to the nearest tenth of a percent.

$\frac{4}{7}$

Note: Rounding to the nearest tenth of a percent requires performing the division to ________ decimal places.

CAUTION!! When is rounding performed during a problem?

Summary:

Fraction	Decimal	Percent
$\frac{32}{100}$		
$\frac{5}{8}$		
$\frac{4}{7}$		

CONCEPT CHECK: If you are asked to solve a problem with the directions "write $\frac{6}{11}$ as a percent, rounded to the nearest tenth of a percent." What is the result before you round?

GUIDED EXAMPLES:

Write the % as a fraction in simplest form.

1. 65%

2. $\frac{1}{4}\%$

Write the fraction as a percent. Round to the nearest tenth of a percent if necessary.

3. $\frac{3}{8}$

Name: ______________________ Date: ______________

Instructor: ______________________ Section: ______________

Notebook 6.7
The Percent Equation

OTHER NOTES

What is the Percent Equation?

What does it look like in equation form?

"What amount is a percent of the whole?

Notes:

Usually "base" appears ….

"is" translates to …

"of" translates to …

Any of the quantities, ________, __________, or ________, can be the unknown quantity

Usually a is used for ____________, p for the ___________, and b for ___________.

In symbols, the percent equation is ____________________.

EXAMPLES 1 & 2 Finding the amount.

What amount is 30% of 140?

60% of 90 is what amount?

EXAMPLES 3 & 4 Finding the base.

34 is 50% of what number?

60% of what number is 42?

EXAMPLE 5 45 is what percent of 180?

EXAMPLE 6 What percent of 30 is 10? Round to the nearest tenth of a percent.

Note: Rounding to the nearest tenth of a percent requires performing the division to _________ decimal places.

CAUTION!! True/False Rounding can be done at any time.

CONCEPT CHECK: For the problem "45 is 50 percent of what number," which quantity is unknown?

GUIDED EXAMPLES:
Use the percent equation to find the following. Round to the nearest tenth of a percent as needed.

1. What amount is 75% of 240?

2. 90 is 15% of what number?

3. 18 is what percent of 60?

4. 20 is what percent of 120?

Name: ______________________________ Date: ________________

Instructor: ___________________________ Section: ______________

Notebook 6.8
The Percent Proportion

OTHER NOTES

What does the percent proportion state?

What is the percent proportion?

Usually "base" appears ….

Any of the quantities, ________, __________, or ________,
can be the unknown quantity

Usually *a* is used for ____________, *p* for the __________,
and *b* for ____________.

Since *p* is divided by 100, do we need to convert from a decimal?

Note: The percent proportion is an alternative to ______________.

EXAMPLE 1 What amount is 20% of 85?

Percent =

Base =

Find the amount:

EXAMPLE 2 21 is 10.5% of what number?

Percent =

Amount =

Find the base:

EXAMPLE 3 83 is what percent of 332?

Amount =

Base =

Find the percent:

EXAMPLE 4 4 is what percent of 48? Round to the nearest tenth of a percent.

Amount =

Base =

Find the percent:

CONCEPT CHECK: What is the unknown quantity in the problem "19 is 17% of what number"?

GUIDED EXAMPLES:

1. What amount is 25% of 64?

2. 25 is 12.5% of what number?

3. 12 is what percent of 60?

4. 30 is what percent of 45? Round to the nearest tenth of a percent.

Name: ______________________________ Date: ______________

Instructor: ____________________________ Section: _____________

Notebook 6.9
Percent Applications

Write the Percent Equation:

Write the Percent Proportion:

Note: Either the percent equation or the percent proportion can be used to solve problems involving ______________.

Note: If you know the percent and the base, which one would be better to use?

Write the rule for finding the amount of sales tax and total cost.
Amount of Sales Tax =

Total Cost =

EXAMPLE 1 Eli purchased a wrist watch for $60. If the sales tax rate is 7%, how much sales tax did Eli pay? What was the total cost of the watch?

Amt of sales tax:

Total cost:

Write down the rule for finding the amount of discount and sale price.
Amount of Discount =

Sale Price =

EXAMPLE 2 Lucky's Shoes is having a 30% discount sale. How much will Jan save on a $40 pair of shoes? What is the sale price of the shoes?

Amt of discount:

Sale price:

Write the rule for finding the percent of change.

Percent of change =

OTHER NOTES

EXAMPLE 3 Enrollment at the local high school went from 320 students to 480 students in one year? Find the percent of increase.

Amt of increase:

Percent change:

What do you know about simple interest?

It is often used in…

It uses an annual rate to calculate ….

It is usually calculated __________, not ____________.

Write down the rule for finding simple interest and total amount paid.

Amount of interest =

Total amount paid =

Notes:

Principal is the amount of the ______________________.

The interest rate is the percent charged per year, expressed as a ___________________.

The amount of time must be expressed in _____________.

EXAMPLE 4 Calculate the amount of interest and the total amount paid for a loan of $2000 for 3 years at a 4% annual rate of interest.

Interest:

Total amount paid:

CONCEPT CHECK: Arty is buying a $50 watch. Sales tax is 9.25%. What is his first step to calculate the total cost of the watch?

GUIDED EXAMPLES:

Alesha purchases a 3-speed bike for $450.

1. If the sales tax is 8%, how much sales tax did Alesha pay?

2. What was the total cost of his bike?

James bought a washer and dryer combination for $1100. He bought it on credit for 1 year at 6% simple interest.

3. Find the amount of interest James will pay.

4. Find the total cost of the washer and dryer, including interest.

Name: ______________________________ Date: ________________

Instructor: ___________________________ Section: ______________

Notebook 7.1

U.S. Length

OTHER NOTES

Length can be measured in various units.
What is one way to measure length?

What determines the unit of length you use?

TABLE 1 U.S. Units of Length	
Unit	**Abbreviation**
inch	
foot	
yard	
mile	

TABLE 2 U.S. Length Conversions
12 in. = 1 ft
36 in. = 1 yd
3 ft = 1 yd
5280 ft = 1 mi
1760 yd = 1mi

_______________________ are used to convert units of measure.

What is a unit fraction?
Give some examples of unit fractions for U.S. length:

What is the rule for finding the appropriate unit fraction?

Write the steps for converting using unit fractions.

1.
2.
3.

EXAMPLE 1 Convert the following using unit fractions. 24 yards to feet

EXAMPLE 2 Convert the following using unit fractions. 15,840 feet to miles

EXAMPLE 3 Convert the following using unit fractions. 4.5 yards to inches

EXAMPLE 4 Convert the following using unit fractions. 90 inches to yards

Write the steps for converting to mixed units.

1.

2.

3.

EXAMPLE 5 Gage is 54 inches tall. Convert this to feet and inches.

CONCEPT CHECK:

When converting 56 inches to feet and inches, what number will represent the number of inches in the final answer?

GUIDED EXAMPLES:

1. Convert using unit fractions. 2.4 miles to yd

2. Convert using unit fractions. 144 in. to yd

3. Convert 102 in. to ft.

4. Convert the height to feet and inches. J.C. is 75 inches tall.

Name: ______________________________ Date: ________________

Instructor: ____________________________ Section: ______________

Notebook 7.2
U.S. Weight and Capacity

OTHER NOTES

What is the difference between weight and mass?

Weight is …

Mass …

Which is used in the U.S. System of Measurement?
Which is used in the Metric System of Measurement?

What determines which unit of weight you use?

TABLE 1 U.S. Units of Weight	
Unit	**Abbreviation**
ounce	
pound	
ton	

TABLE 2 U.S. Weight Conversions
16 oz = 1 lb
2000 lb = 1 ton

Give some examples of unit fractions for U.S. units of weight.

What is the rule for finding the appropriate unit fraction?

Write the steps for converting using unit fractions.
1.
2.
3.

EXAMPLE 1 Convert the following using unit fractions. 7.5 tons to pounds

EXAMPLE 2 Convert the following using unit fractions. 64 ounces to pounds

Give some examples of unit fractions for U.S. units of capacity.

EXAMPLE 3 Convert the following using unit fractions. 7.5 pints to cups

EXAMPLE 4 Convert the following using unit fractions. 26 quarts to gallons

Write the steps for converting to mixed units.

1.

2.

3.

EXAMPLE 5 LaShonda buys a bottle of ketchup that contains 44 fl oz. Convert this to pints and fluid ounces.

CONCEPT CHECK:

Which unit would be the most appropriate to measure the weight of a cement truck?

GUIDED EXAMPLES:

1. Convert 5000 pounds to tons.

2. Convert 9 lb to oz.

3. Convert 52 qt to gallons.

4. Convert 38 fluid ounces to ________ pints ________ ounces.

Name: ______________________________ Date: ________________

Instructor: ___________________________ Section: ______________

Notebook 7.3
Metric Length

OTHER NOTES

In the United States, what units of measurement do we use for length?

In the metric system, what is the basic unit of length?
All metric units of length are based on the ______________.

How long is a meter?

Which metric units are shorter than the meter?
Which are longer than the meter?

Metric Units of Length references:
Which unit is about as thick as a larger paper clip wire?
Which unit is about the width of your pinky finger?
Which unit is about the width of your hand?
Which unit is slightly longer than a yard?
Which unit is a little more than 30 feet?
Which unit is about the length of a football field?
Which unit is about 0.6 of a mile?

What does each metric prefix mean?

milli: centi: deci:

deka: hecto: kilo:

Recall: A unit fraction is a fraction whose value is ________.

Write the rule of finding the appropriate unit fraction:

EXAMPLE 1 Convert the following. 400 centimeters to meters

What phrase can help you to remember the order of the metric prefixes?

Note: The larger units are on the _______; the smaller units on the ________.
When converting larger units to smaller units, which direction do you move?
When converting smaller units to larger units, which direction do you move?

EXAMPLES 2 & 3 Convert the following.

87.45 millimeters to decimeters

How many places do you move?

Which direction do you move?

96.3 kilometers to meters

How many places do you move?

Which direction do you move?

EXAMPLE 3 Convert the following. 150 dekameters to centimeters

How many places do you move?

Which direction do you move?

CONCEPT CHECK:

When converting 2 dekameters to millimeters, how many places should the decimal point be moved and in which direction?

GUIDED EXAMPLES:

1. Convert 45.1 m to km.
2. Convert 3.1 hm to cm.
3. Convert 46 cm to mm.
4. Convert 25 cm to meters.

Name: ______________________________ Date: ______________

Instructor: ____________________________ Section: ______________

Notebook 7.4
Metric Mass and Capacity

OTHER NOTES

In the United States, what units of measurement do we use for weight?

In the metric system, what is the basic unit of mass?
All metric units of mass are based on the _______________.

How much is a gram?

How much is a kilogram?

Which metric units are smaller than the gram?
Which are larger than the gram?

Recall each metric prefix:

milli:	centi:	deci:
deka:	hecto:	kilo:

Recall: A unit fraction is a fraction whose value is _________.

EXAMPLE 1 Convert the following. 32 centigrams to grams

What phrase can help you to remember the order of the metric prefixes?

Note: The decimal point in the number moves the __________ of places and in the ____________________ as the number of moves in the list.

EXAMPLES 2 & 3 Convert the following.
123.5 milligrams to grams
How many places do you move?
Which direction do you move?

216 kilograms to centigrams
How many places do you move?
Which direction do you move?

In the United States, what units of measurement do we use to measure liquid capacity?

In the metric system, what is the basic unit of capacity?
All metric units of capacity are based on the _______________.

How much is a liter?

How much is a kiloliter?

Which metric units are smaller than the liter?
Which are greater than the liter?

How much is one milliliter?

EXAMPLES 4 & 5 Convert the following.
900 milliliters to liters
How many places do you move?
Which direction do you move?

83.2 liters to centiliters
How many places do you move?
Which direction do you move?

CONCEPT CHECK:
Which is the most appropriate unit to measure the capacity of a pool?

GUIDED EXAMPLES:
1. Convert 4.08 mg to dg.

2. Convert 9.332 kg to cg.

3. Convert 4.03 mL to cL.

4. Convert 5.2 L to cc.

Name: ______________________ Date: ____________

Instructor: ____________________ Section: ____________

Notebook 7.5
Converting Between U.S. and Metric Units

OTHER NOTES

Write down the unit fractions for converting lengths.

U.S. to Metric (Length)

Recall the fraction used to make conversions:

Metric to U.S. (Length)

EXAMPLE 1 Convert the following. 44.02 miles to kilometers

Conversion:

EXAMPLE 2 The bases in a major league baseball stadium are 90 feet apart. In terms of meters, how far apart are the bases?

Conversion:

Write down the unit fractions for converting mass.

U.S to Metric (Weight/mass)

Metric to U.S. (Weight/mass)

EXAMPLE 3 Convert the following. 5.2 kilograms to pounds

Conversion:

EXAMPLE 4 Convert the following. 6 pounds to grams
Conversion:

Write down the unit fractions for converting capacity.
U.S to Metric (Capacity)

Metric to U.S. (Capacity)

EXAMPLE 5 Convert the following. 6 pounds to grams
Conversion:

EXAMPLE 6 Convert the following. 3 quarts to liters
Conversion:

CONCEPT CHECK:
A recipe lists the number of grams needed for an ingredient, but Sarah only has a scale that uses U.S. measurements. She will most likely convert the amount from the recipe to which unit?

GUIDED EXAMPLES:
1. Convert 225.6 m to yd.

2. Convert 2.6 in. to yd. Round to the nearest hundredth.

3. Convert 12.6 gal to L.

4. Convert 16.45 L to gal. Round to the nearest tenth.

Name: ______________________________ Date: ________________

Instructor: ____________________________ Section: ______________

Notebook 7.6
Time and Temperature

Write down the unit fractions for converting times.

Recall the fraction used to make conversions:

EXAMPLE 1 Convert 195 minutes to hours.

EXAMPLE 2 Convert 4.4 hours to seconds.

What are the two scales for temperature?

What is the basic unit for each scale?

What are some common temperature equivalencies?

Water freezes:

Room temperature:

Water boils:

What is the formula used to convert from Fahrenheit to Celsius?

EXAMPLE 3 Convert 36° Fahrenheit to Celsius. Round to the nearest tenth.

OTHER NOTES

What formula is used to convert from Celsius to Fahrenheit?

EXAMPLE 4 Convert 81° Celsius to Fahrenheit. Round to the nearest tenth.

EXAMPLE 5 Marquez, who lives in the United States, purchased a heat pump that was made in Belgium. The heat pump can operate efficiently up to a temperature of 35° Celsius. To the nearest degree, what is this temperature in Fahrenheit?

Understand the problem

Create a plan

Find the answer

Check the answer

CONCEPT CHECK:

What is the unit fraction needed to convert seconds to minutes?

GUIDED EXAMPLES:

1. Convert 5 hours to minutes using unit fractions.

2. Convert 600 hours to days.

3. Convert 60° F to C.

4. Convert 10.3° C to F.

Name: ______________________________ Date: ________________

Instructor: ___________________________ Section: ______________

Notebook 8.1
Lines and Angles

What is a point?
How is a point represented?
Ex.

What is a line?
How is a line named?
Ex.

What is a line segment?
How is a line segment named?
Ex.

What is a ray?
How is a ray named?
Ex.

What is an angle?
How is an angle named?
Ex.

EXAMPLES 1–3 Identify each of the following figures as a line, a line segment, or a ray. Give the name of each.

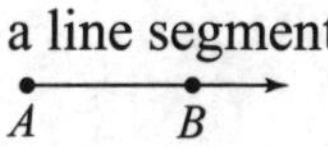

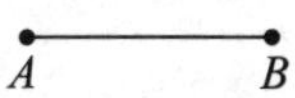

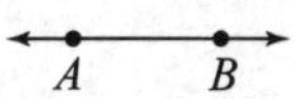

How is an angle formed?
An angle is measured by ______________________________.
How is an opening measured?

Define four types of angles and give an example of each.
1.

2.

3.

4.

OTHER NOTES

EXAMPLES 4 & 5 Identify each angle below as straight or right.

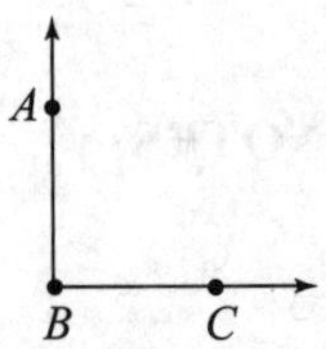

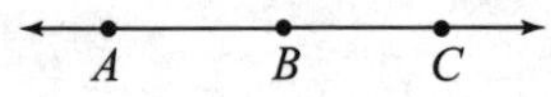

Identify each angle below as either acute or obtuse.

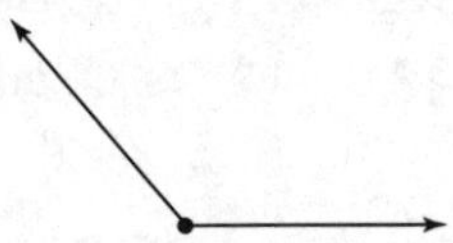

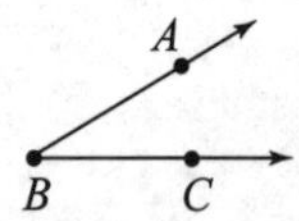

Two angles are _____________ if the sum of their measures is ______.
Give an example.

Two angles are ______________ if the sum of their measure is ______.
Give an example.

EXAMPLES 6 & 7 Identify each angle pair as complementary, supplementary, or neither. You will need to draw the angles.

EX. EX.

Find the complement of a 65° angle.

Find the supplement of a 75° angle.

CONCEPT CHECK:

How do you find the complement of an angle?

GUIDED EXAMPLES:

1. Identify the figure as a line, a line segment, or a ray.

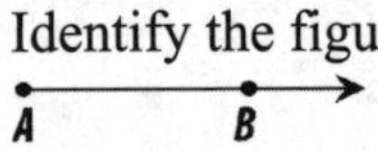

2. Identify the angle as acute, obtuse, or neither.

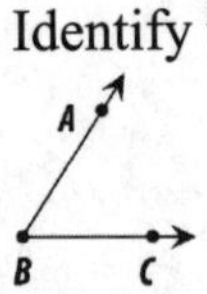

3. Identify the angle pair as complementary, supplementary, or neither. angles of 54° and 25°

4. Find the measure of the complement of a 23° angle.

Name: ______________________________ Date: ______________

Instructor: ____________________________ Section: ____________

Notebook 8.2
Figures

OTHER NOTES

What is a polygon?

What does it mean that a line is one-dimensional?

Why are polygons considered two-dimensional?

A figure can be classified by the number of __________.
A 3 sided figure is a ____________________.

A 4 sided figure is a ____________________.

TRIANGLES–Draw an example of each.
What kind of triangle has all three sides equal in length?

What kind of triangle has two sides equal in length?

What kind of triangle has all three sides of different lengths?

Property: The sum of the measures of the three angles of a triangle is ________.

A triangle can also be classified according to the ________________.
A triangle with a 90 angle is a ______________________.

A triangle with three acute angles is an ______________________.

A triangle with an obtuse angle is an ______________________.

EXAMPLE 1 If two angles of a triangle measure 55° and 70°, find the measure of the third angle. Then, identify the triangle as acute, right, or obtuse.

QUADRILATERALS–Draw an example of each.

Rectangle Square Trapezoid

Parallelogram Diamond Kite

A ____________ is a quadrilateral with opposite sides that are ____________ in length and four angles that are ________________.

Ex.

A square is a quadrilateral with ________________ that are ______ in length and__________ angles that are ______________________.

Ex.

What are parallel lines?

What is the symbol for parallel lines?

Ex.

A parallelogram is a __________________, or four sided figure, with opposite sides that are ___________ and ______________.

What is the difference between a rectangle and a parallelogram?

A ______________ is a quadrilateral with only ______ pair of __________________ that are parallel.

CONCEPT CHECK:

A triangle has angles that measure 57°, 72°, and 51°. What kind of triangle is it?

GUIDED EXAMPLES:

1. Find the measure of the third angle in each triangle.
   triangle with angles of 27° and 32°

2. Identify the triangle as acute, right, or obtuse.
   a triangle with angles of 55°, 75°, and 50°

3. Identify each triangle as acute, right, or obtuse.
   a triangle with angles of 45°, 45°, and 90°

4. Identify the figure below.

Name: ______________________________ Date: ________________

Instructor: ____________________________ Section: ______________

Notebook 8.3
Perimeter—Definitions and Units

OTHER NOTES

What is a polygon?

Define perimeter.

How do you find the perimeter of a polygon?

Perimeter always represents ____________ and is measured in units of __________, such as …

Which of these situations, if any, would involve perimeter?
- Painting a room
- Spreading grass seeds
- Hanging gutters around a house

Which of these measurements, if any, could represent perimeter?
- 19 inches
- 34 cups
- 2.3 square feet

EXAMPLE 1 Find the perimeter of the figure.
You will need to draw the triangle.

EXAMPLE 2 Find the perimeter of the figure.
You will need to draw the figure.

EXAMPLE 3 Find the perimeter of the figure.
You will need to draw the figure.

EXAMPLE 4 Find the distance around the infield of a major league baseball field. You will need to draw the figure.

CONCEPT CHECK:

A polygon is a closed figure with __________ or more sides.

GUIDED EXAMPLES:

1. Find the perimeter.

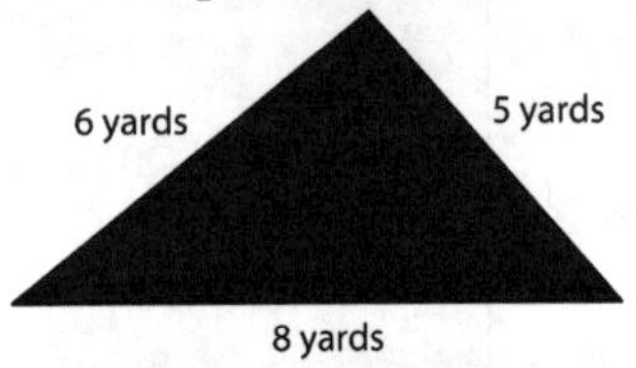

2. Find the perimeter.

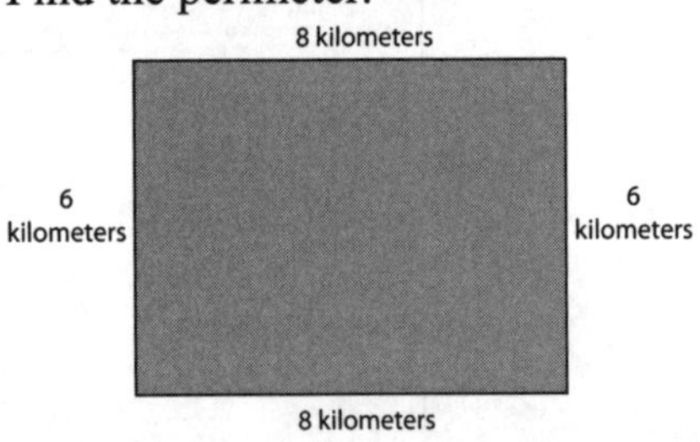

3. Tennessee is shaped like a parallelogram. Find the approximate length of the border around the state using the following figure.

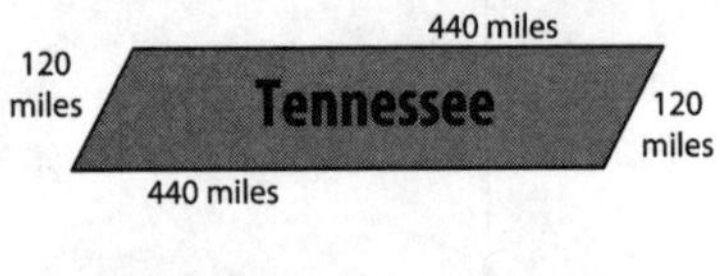

4. Find the perimeter of the figure.

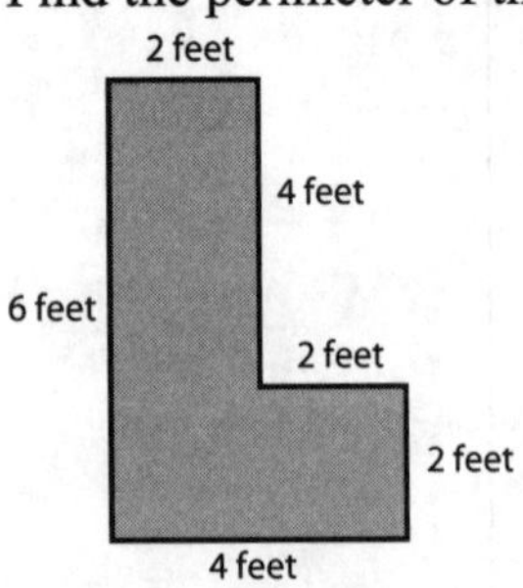

Name: ______________________________ Date: ________________

Instructor: ____________________________ Section: ______________

Notebook 8.4
Finding Perimeter

OTHER NOTES

To find the perimeter of a figure, _________________________.
Perimeter is measured in units of ____________.
A ________________ is a three-sided geometric figure.
A ________________ is a quadrilateral with opposite sides that are ____________ in length and four angles that are __________.
A ________________ is a quadrilateral with four sides that are ____________ in length and four angles that are ____________.

This video focuses on learning ___________ for finding _____________.

What is the formula for finding the perimeter of a triangle?
Ex.

EXAMPLE 1 Use the formula for the perimeter of a triangle to find the perimeter of a triangle with side lengths of 3.5 cm, 2.5 cm, and 4 cm.

What is the formula for finding the perimeter of a rectangle?
Ex.

EXAMPLE 2 Use the formula for the perimeter of a rectangle to find the perimeter of a rectangular field with a length of 40 feet and a width of 30 feet.

What is the formula for finding the perimeter of a square?
Ex.

EXAMPLE 3 Find the perimeter of a square to find the perimeter of a square court with side lengths of 14 meters.

EXAMPLES 4–6 For each figure, find the perimeter.

A triangular garden with side lengths of 5 feet, 7 feet and 10 feet.

A rectangular photo with length 20.5 cm and a width 10 cm.

A square tabletop with side lengths of $2\frac{1}{2}$ m.

CONCEPT CHECK:

To find the perimeter of a triangle, how many side lengths must be known?

GUIDED EXAMPLES:

Use the appropriate perimeter formula to find the perimeter of the figures shown.

1.

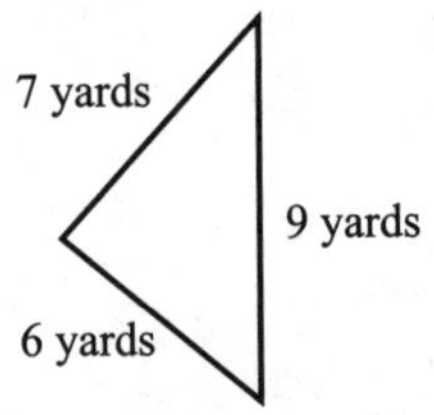

2.

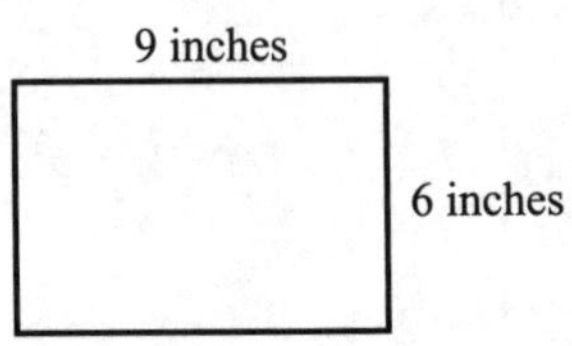

3.

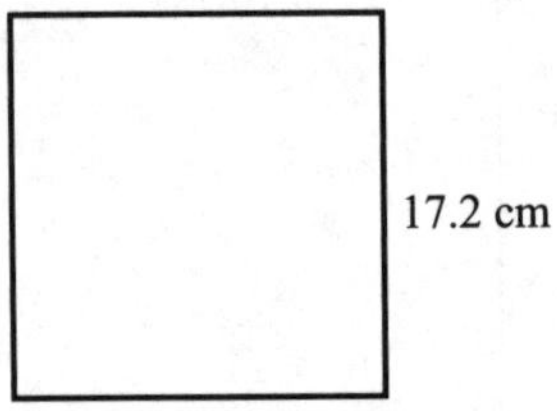

Name: ______________________ Date: ____________

Instructor: ______________________ Section: ____________

Notebook 8.5
Area—Definitions and Units

OTHER NOTES

How do you find the area of a figure?

Area is measured in ______________________, such as …

What does one square unit look like?

Square inch:

Square centimeter:

Which of these situations, if any, would involve area?

Laying a carpet
Putting up a fence
Filling a swimming pool

Which of these units, if any, could represent area?

36 inches
6 gallons
14 square feet

EXAMPLES 1 & 2 How many square units are needed to cover each figure completely? You will need to draw the figures.

EXAMPLE 3 Draw two different rectangles, each of which has an area of 6 square units.

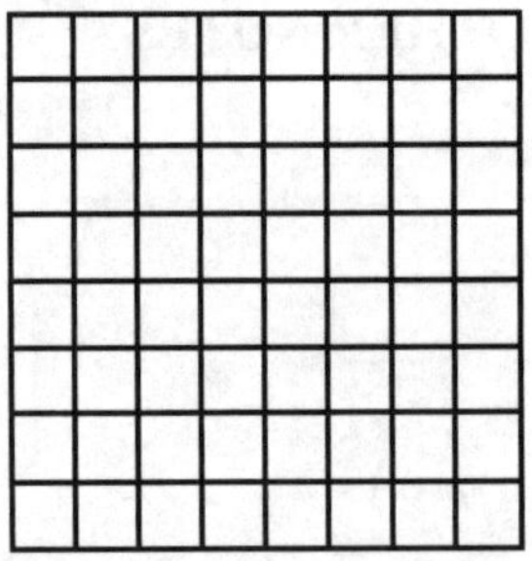

EXAMPLE 4 Find the area of a square that measures 5 cm on each side.

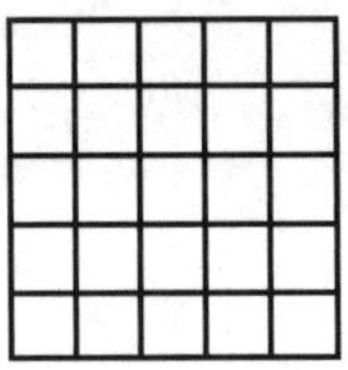

EXAMPLE 5 Find the area of a right triangle with a base and height of 4 in.

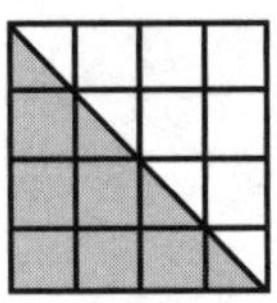

CONCEPT CHECK: A triangle covers 9 whole squares and 5 half squares. What would this area be?

GUIDED EXAMPLES:

1. Find the area of a right triangle with a base and height of 3 in.

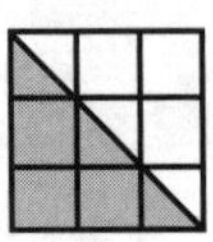

2. Draw a rectangle which has an area of 12 square units.

3. Which of these situations, if any, would involve area?
    a. Measuring flour for a cake
    b. Painting a room
    c. Framing a picture
    d. Laying a tile floor

Name: ______________________________ Date: ________________

Instructor: ___________________________ Section: ______________

Notebook 8.6
Finding Area

OTHER NOTES

Area is the measure of the __________ inside a figure.
To find the area, ______________________________.
Area is measured in ____________________________.
A ________________ is a three-sided geometric figure.
A ________________ is a quadrilateral with opposite sides that are ___________ in length and four angles that are __________.
A ________________ is a quadrilateral with four sides that are ____________ in length and four angles that are __________.

This video focuses on learning __________ for finding ____________.

What is the formula for finding the area of a rectangle?
Ex.

EXAMPLE 1 Use the formula for the area of a rectangle to find the area of the photo. You will need to draw the rectangle.

What is the formula for finding the area of a square?
Ex.

EXAMPLE 2 Use the formula for the area of a square to find the area of a square deck with side lengths of 3 meters.

What is the formula for finding the perimeter of a triangle?
How does the height vary with different types of triangles?
Ex.

EXAMPLE 3 Use the formula for the area of a triangle to find the area of a right triangle with a base of 4 cm and a height of 3 cm.

EXAMPLE 4 Find the area of the infield of a major league baseball diamond, which is a square whose sides are 90 feet long.
You will need to draw the diagram.

CONCEPT CHECK:
An 8 in. by 11 in. piece of graph paper has a grid with lines that are one inch apart. In other words, it is covered in squares that measure 1 inch on each side. How many squares, or square units, cover the graph paper?

GUIDED EXAMPLES:
Find the area of the following.

1. A rectangle whose length is 15 inches and width is 4 inches.

2. A square whose sides are 6 cm long.

3. A right triangle with a base of 6 cm and a height of 5 cm.

4. A rectangular patch of wild flowers which has a length of 4.5 yards and a width of 3.2 yards.

Name: ______________________ Date: ______________

Instructor: ______________________ Section: ______________

Notebook 8.7
Understanding Circles

OTHER NOTES

Define a circle.

What is the difference between a radius and a diameter?

radius—distance from _________ to ______________.

diameter—distance ______________________________.

Label each part of a circle.

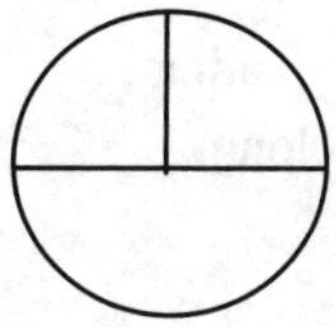

What is the relationship between the diameter and the radius?

What are the formulas for diameter and radius?

EXAMPLE 1 Use the appropriate formula to find the radius of a pizza with a diameter of 10 inches.

EXAMPLE 2 Which situation involves diameter and which involves radius?

A neighbor purchased a 14-foot circular trampoline for his children.

In softball, the circle around the pitcher's mound is drawn 6 feet from where the pitcher stands.

What is circumference?

What is pi?
Ex.

What are the two approximations we can use for π?

CONCEPT CHECK:
A wheel makes one revolution on the ground. What part of the circle can be used to determine the distance traveled?

GUIDED EXAMPLES:

1. Does the situation described involve the diameter of the radius?
   A toy helicopter has moveable blades that are 5 inches long.

2. Does the situation described involve the diameter of the radius?
   A child's circular board game folds along the center with a measure of 15 inches.

3. Find the radius of a circular garden with a diameter of 3.4 feet.

4. Find the radius of a circular drum with a diameter of 25 cm.

Name: ______________________________ Date: ______________

Instructor: ___________________________ Section: ____________

Notebook 8.8
Finding Circumference

OTHER NOTES

The _____________ is the distance around the circle.
What is π?

The formula for finding the circumference of a circle is different depending on what is given.
What is the formula for finding the circumference of a circle,
… when you are given the radius?

…when you are given the diameter?

What are the units for circumference?

Exact answers for circumference are in terms of __________.
Approximate answers can be obtained using _____ or _______ for π.

EXAMPLE 1 Find the circumference of a circle with a radius of 8 inches. Find the exact answer in terms of π. Then find the approximate value using 3.14 for π.

EXAMPLE 2 Find the circumference of a circle with a diameter of 5 centimeters. Find the exact answer in terms of π. Then find the approximate value using 3.14 for π.

EXAMPLE 3 Find the circumference of a circle with a diameter of 21 yards. Use the approximate value of $\frac{22}{7}$ for π.

EXAMPLE 4 Find the circumference of a circle with a radius of 3.5 miles. Use the approximate value of 3.14 for π.

EXAMPLE 5 The earth's equator forms a circle. Estimate the number of miles a ship would have to travel if it went around the earth at the equator. Use 7900 miles as an approximation of the earth's diameter. Use 3.14 for π.
Understand the problem

Create a plan

Find the answer

Check the answer

CONCEPT CHECK:
A circle has a diameter of 10 feet. What is the first calculation that would be performed to find the circumference. If needed, use 3.14 for π.

GUIDED EXAMPLES:
Find the circumference of the following circles. Find the exact answer in terms of π. Then find the approximation of 3.14 for π. Round to the nearest tenth.

1. A circle with diameter of 3 feet.

2. A circle with a radius of 4 inches.

Find the circumference of each circle with the given diameter. Use $\frac{22}{7}$ for π.

3. 21 cm

4. 14 in.

Name: ______________________________ Date: ________________

Instructor: ____________________________ Section: ______________

Notebook 8.9
Finding Area—Circles

OTHER NOTES

The area of a figure is the measure of the __________ inside the figure.

Area is measured in ______________ of length.

What is the formula for finding the area of a circle with radius, r?

That is, to find the area, multiply _____ by ______________.

What are the steps for finding the area of a circle?

1. Find the _____________.
2.
3.

EXAMPLE 1 Find the area of a circle with a radius of 6 feet. Use the approximate value of 3.14 for π.

EXAMPLE 2 Find the area of a circle with a diameter of 6 meters. Use the approximate value of 3.14 for π.

EXAMPLE 3 Find the area of a circle with a diameter of 7 feet. Use the approximate value of $\frac{22}{7}$ for π.

EXAMPLE 4 The diameter of a quarter is 2.5 cm. Find the area of a quarter, rounded to the nearest hundredth. Use the approximate value of 3.14 for π.
Understand the problem

Create a plan

Find the answer

Check the answer

CONCEPT
What is the formula for finding the area of a circle?

GUIDED EXAMPLES:
Find the area of the following circles.
1. A circle with a radius of 12 feet. Use 3.14 for π.

2. A circle with a diameter of 28 cm. Use $\frac{22}{7}$ for π.

3. A circle with a radius of 0.3 meters. Use 3.14 for π. Round to the nearest thousandth.

4. A circle with a diameter of 4 ft. Use 3.14 for π.

Name: ______________________________ Date: ________________

Instructor: ___________________________ Section: ______________

Notebook 9.1
Volume Definitions and Units

OTHER NOTES

The distance around the figure is the ______________.
Perimeter is measured in units of ________________.
The measure of the surface inside the figure is the __________.
Area is measured in ________________________________.

Complete the following table:

Measurements and Dimensions			
Measurement	Number of Dimensions	Dimensions	Label in Feet

Note: The exponent on the unit is equal to ____________________.

What is volume?
How do we measure volume?
Consider this: "How many __________ does it take to fill the object completely?"

Define rectangular box.

Define cube.

What is the volume of a cube that measures 1 inch on each side?
What is the volume of a cube that measures 1 centimeter on each side?

Which of these situations, if any would involve volume?
- Tiles on a floor
- Fencing around a garden
- Water inside a barrel

Which of these units, if any, could represent volume?
- 65 miles
- 6 cubic yards
- 12 square inches

Write the steps for finding the volume of a box by counting cubes.
1.

2.

3.

EXAMPLE 1 Find the volume of a box with a length of 4 inches, a width of 3 inches, and a height of 4 inches. You will need to draw the picture.

EXAMPLE 2 Find the volume of a cube that measures 4 cm on each side. You will need to draw the picture.

EXAMPLE 3 Find the volume of a box with length 5 m, width 1 m, and height 3 m. Draw the picture.

CONCEPT CHECK:
Volume involves working in how many dimensions?

GUIDED EXAMPLES:

1. Find the volume of a rectangular box with length 2 ft, width 4 ft, and height 3 ft.

2. Find the volume of a cube with the given side measure 3 cm.

3. Find the volume of a rectangular box with length 5 in., width 2 in., and height 4 in.

Name: ______________________ Date: ______________

Instructor: ______________________ Section: ______________

Notebook 9.2
Finding Volume

OTHER NOTES

The amount of space inside a figure is the ______________.
Volume is measured in ______________.
A ______________ is a three-dimensional figure with 6 sides, and each side is a ______________.
A __________ is a rectangular box in which every side is a ______________.

We will learn formulas for finding volume of a rectangular box, a cube, a cylinder, and a sphere.

What is the formula for finding the volume of a rectangular box?
You will need to draw the box and label the sides.

EXAMPLE 1 Use the formula to find the volume of a rectangular box with a length of 12 inches, a width of 10 inches, and a height of 5 inches. You will need to draw and label the box.

What is the formula for finding the volume of a cube?
You will need to draw the cube and label the sides.

EXAMPLE 2 Use the formula to find the volume of a cube that measures 2.5 km on each side. You will need to draw and label the box. Round to the nearest hundredth.

What is the formula for finding the volume of a cylinder?
You will need to draw the cylinder and label the height and radius.

EXAMPLE 3 Use the formula and the approximate value of 3.14 for π to find the volume of a cylinder with a radius of 3 inches and a height of 5 inches. Draw your cylinder and label the radius and height.

What is the formula for finding the volume of a sphere?
You will need to draw the sphere and label the radius.

EXAMPLE 4 Use the formula and the approximate value of 3.14 for π to find the volume of a sphere with a radius of 6 mm. Draw your sphere and label the radius.

EXAMPLE 5 Find the volume of a soda can (in cubic centimeters) with a diameter of 6.6 cm and a height of 12.1 cm. Use the approximate value of 3.14 for π. Round to the nearest tenth.

Understand the problem

Create a plan

Find the answer

Check the answer

CONCEPT CHECK: The formula $V = \pi r^2 h$ is used to find the volume of what object?

GUIDED EXAMPLES:

1. Find the volume of a rectangular box with length of 4.5 ft, width of 3 ft and a height of 6 ft.

2. Find the volume of a cube with a side length of 7 inches.

3. Find the volume of a cylinder with a radius of 5 cm, and a height of 4 cm. Use 3.14 for π.

4. Find the volume of a sphere with a radius of 3 mm. Use 3.14 for π.

Name: ______________________________ Date: ________________

Instructor: ____________________________ Section: ______________

Notebook 9.3
Square Roots

OTHER NOTES

What does it mean if a number is squared?

What is a perfect square?

List the first 15 perfect squares:

Is 81 a perfect square? Why or why not?

Is 50 a perfect square? Why or why not?

Is 64 a perfect square? Why or why not?

Is 27 a perfect square? Why or why not?

When you multiply two identical factors to result in another number, the ________________ of the number is one of those _____________.

Finding a square root is the reverse operation of ___________________.

What is the symbol for finding the square root?
What is this symbol called?

Define square root.

EXAMPLE 1 Find the square root of 36.

EXAMPLE 2 Simplify $\sqrt{64}$.

EXAMPLE 3 Simplify $\sqrt{400}$.

EXAMPLE 4 The area of a square is 49 cm^2.
What is the length of each side?

EXAMPLE 5 Use a calculator to approximate $\sqrt{72}$. Round to the nearest hundredth.

What is another way to approximate the square root of a number?

EXAMPLE 6 Approximate $\sqrt{75}$ by finding the two consecutive whole numbers between which the square root lies.

CONCEPT CHECK:

To approximate $\sqrt{34}$, between which two consecutive whole numbers would the square root be?

GUIDED EXAMPLES:

1. Identify whether the following is a perfect square. 39

2. Simplify the square root. $\sqrt{121}$

3. Use a calculator to approximate each square root. Round to the nearest hundredth. $\sqrt{200}$

4. Find the two consecutive whole numbers between which the square root lies. $\sqrt{51}$

Name: ______________________________ Date: ________________

Instructor: ___________________________ Section: ______________

Notebook 9.4
The Pythagorean Theorem

OTHER NOTES

How many right angles are in a right triangle?

What are the legs of a right triangle?
Which side is the hypotenuse?
Label each side in the triangle.

State the Pythagorean Theorem.

What is the Pythagorean Theorem used for?

Write the steps for finding the missing side in a right triangle.

1. Identify the _________, the ____________ and the ____________.
To find the hypotenuse, use …
To find one of the legs, use …

2.

3.

EXAMPLE 1 Find the length of the hypotenuse of a right triangle with legs that measure 3 cm and 4 cm.

EXAMPLE 2 One leg of a right triangle measures 9 inches and the hypotenuse measures 15 inches. Find the length of the other leg.

EXAMPLE 3 Find the length of the missing side of the right triangle. Round to the nearest tenth.

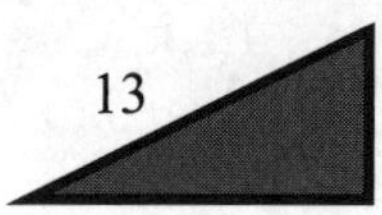

EXAMPLE 4 A standard computer monitor has a length of 20 inches and a width of 15 inches. What is the measure of the diagonal of the computer monitor?

Understand the problem

Create a plan

Find the answer

Check the answer

CONCEPT CHECK:

The Pythagorean Theorem is used to find the length of an unknown length of a side of what kind of triangle?

GUIDED EXAMPLES:

1. A right triangle has legs of lengths 5 feet and 12 feet. Find the length of the hypotenuse.

2. A right triangle has a leg of length 6 inches and a hypotenuse of length 10 inches. Find the length of the other leg.

3. A right triangle has a leg of length 5 m and a hypotenuse of length 7 m. Find the length of the other leg. Round to the nearest tenth.

4. Jacqueline's triangular flower garden has one side along the house and the other by the driveway, meeting the house at a right angle. The house side of the garden is 12 feet long, and the driveway side is 9 feet long. How much edging material will she need for the side that runs from the house to the driveway, which is the hypotenuse of the right triangle?

Name: ______________________________ Date: ________________

Instructor: ____________________________ Section: ______________

Notebook 9.5
Similar Figures

OTHER NOTES

What does it mean if two figures are similar?
Draw similar figures.

True/False Similar figures must have the same shape and the same size.

A _____________ is a statement that two ratios are ______.
What does a proportion look like? _________ = __________.

How do you determine if two figures are similar?

For similar figures, the __________ of corresponding sides are ______________, their lengths have the same ___________.

EXAMPLE 1 Identify the corresponding sides and set up proportions to determine if the figures are similar. You will need to label the sides of the triangles.

EXAMPLE 2 Identify the corresponding sides and set up proportions to determine if the figures are similar.

All squares are _________________.

EXAMPLE 3 Determine if the figures are similar.

EXAMPLES 4 & 5 Determine if the figures in each pair are similar. You will need to label the sides.

CONCEPT CHECK:
What does it mean when "the lengths of corresponding sides of similar figures are proportional"?

GUIDED EXAMPLES: Similar Figures

1. Determine if the figures are similar.

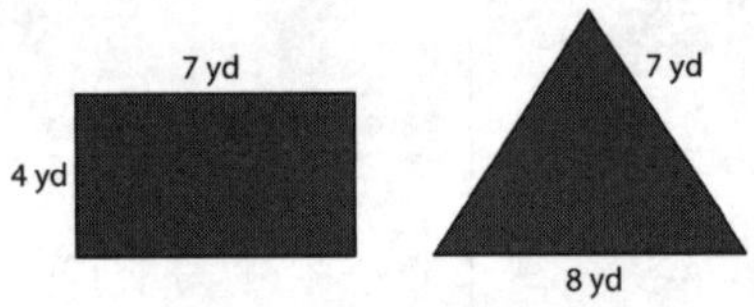

2. Identify corresponding sides in the similar triangles below.

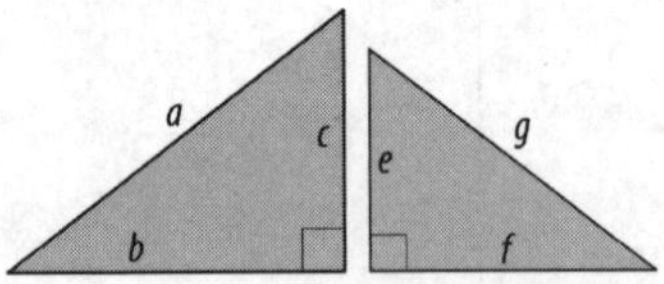

3. Determine if the figures are similar.

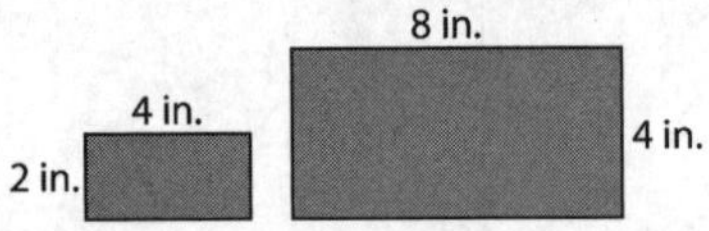

4. Determine if the figures are similar.

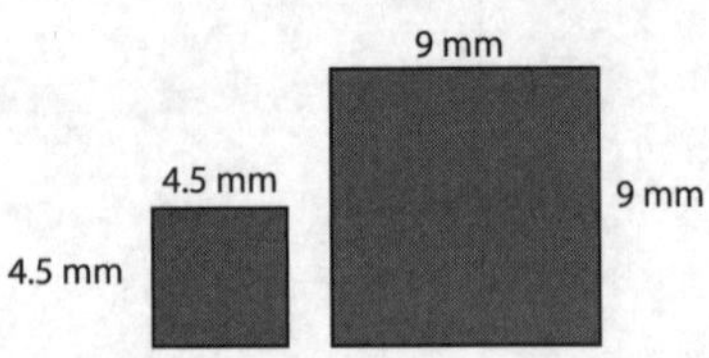

Name: ______________________________ Date: ______________

Instructor: ___________________________ Section: ______________

Notebook 9.6
Finding Missing Lengths

OTHER NOTES

What does it mean if two figures are similar?

Corresponding sides of __________ figures are __________.

EXAMPLES 1 & 2 Write a proportion for the corresponding sides of the similar figures. You will need to label the sides.

Write the steps for finding the unknown side lengths in similar figures.

1.

2.

EXAMPLE 3 Find the unknown side length x in the similar figures. You will need to label the sides.

EXAMPLE 4 Find the unknown side length x in the similar figures.

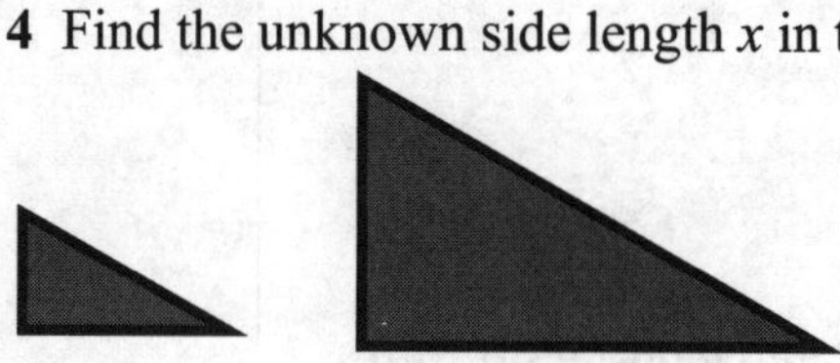

EXAMPLE 5 Find the unknown side length x in the similar figures.
You will need to label the sides.

EXAMPLE 6 Kristen takes a 4-inch wide by 6-inch tall 4" x 6" photo of her family to a photography lab to be enlarged into a poster that will be 48 inches tall. Find the smaller dimension, or width, of the poster.
Understand the problem

Create a plan

Find the answer

Check the answer

CONCEPT CHECK:
Chef Aiko has two similar rectangular cutting boards. One has a length of 11 inches and a width of 9 inches. The other has a width of 6 inches. Set up the proportion to find the unknown length of the second cutting board, x.

GUIDED EXAMPLES:

1. Find the unknown side length in the similar figures

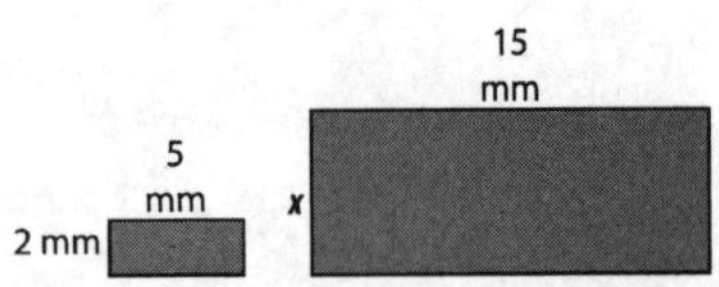

2. Find the unknown side length in the similar figures.

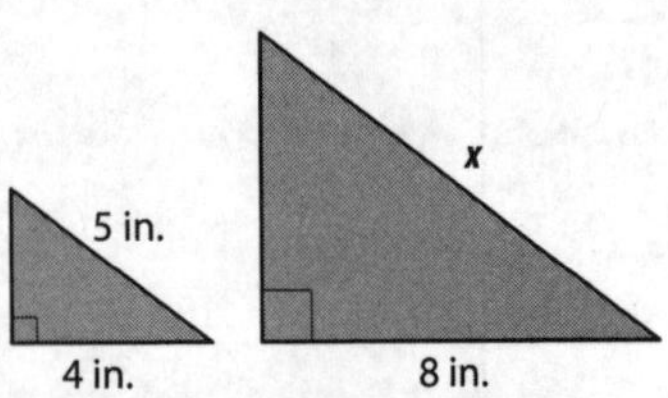

3. Find the unknown side length in the similar figures.

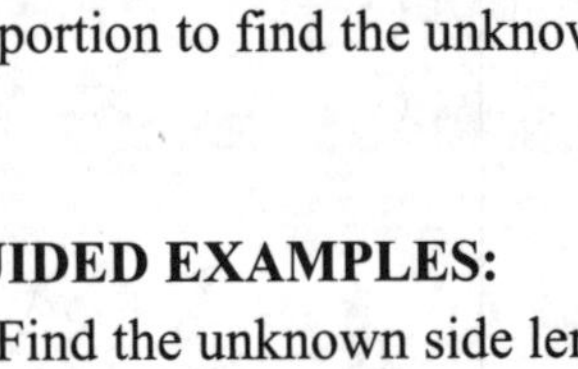

4. Marian took a 3" × 5" photo to a shop to be enlarged into a poster 40" tall. Find the smaller dimension, or width, of the poster.

Name: ______________________ Date: ____________

Instructor: ______________________ Section: ____________

Notebook 9.7
Congruent Triangles

OTHER NOTES

What does it mean if two objects are congruent?

Note. Two triangles are ________ if the measures of corresponding ____________ are equal and the __________ of corresponding sides are the __________.

State the rule for congruence by side-side-side (SSS).

EXAMPLE 1 Explain why $\triangle ABC$ and $\triangle EFD$ are congruent.

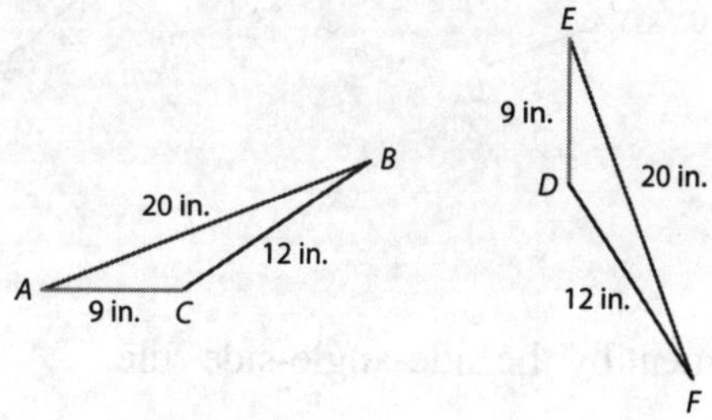

State the rule for congruence by side-angle-side (SAS).
Two triangles are congruent by SAS if the following conditions are met:
1.

2.

EXAMPLE 2 Explain why $\triangle STU$ and $\triangle CAB$ are congruent.

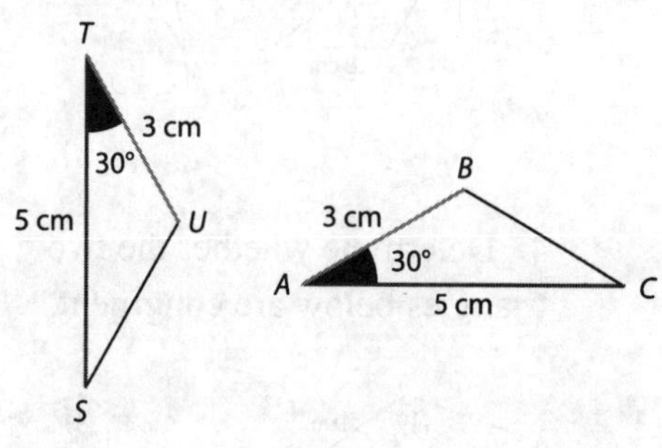

State the rule for congruence by angle-side-angle (ASA).
Two angles are congruent by ASA if the following conditions are met:
1.

2.

EXAMPLE 3 Explain why $\triangle ABC$ and $\triangle STU$ are congruent.

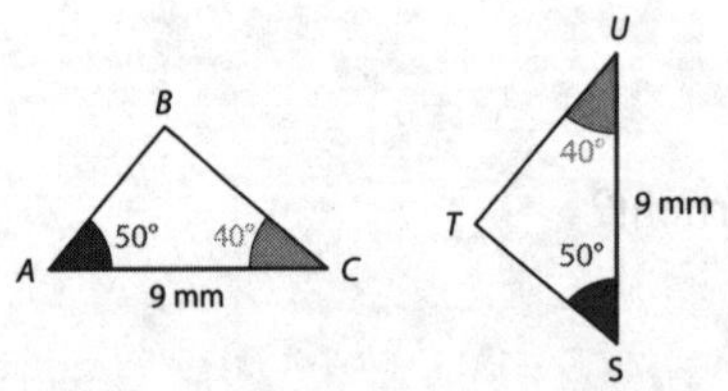

Do all combinations of congruent sides and angles show congruent? Tell why or why not.

EXAMPLES 4 & 5 Determine whether each pair of triangles is congruent.

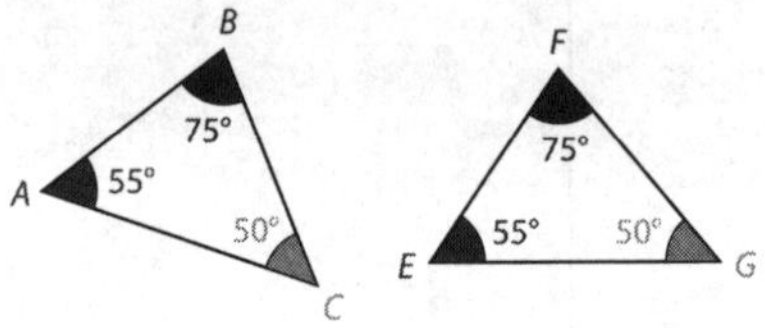

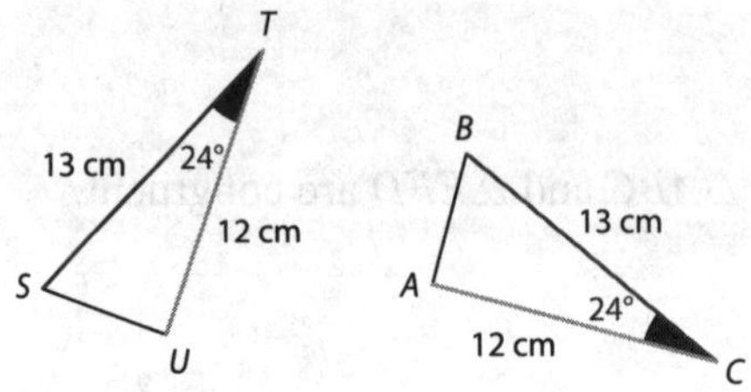

CONCEPT CHECK:

Describe two triangles that are congruent by the side-angle-side rule.

GUIDED EXAMPLES:

1. Explain why the two triangles below are congruent.

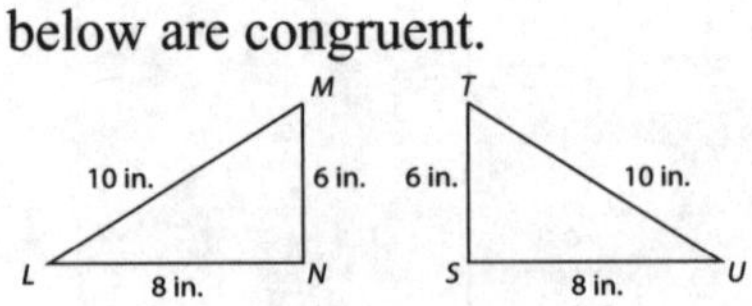

2. Explain why the two triangles below are congruent.

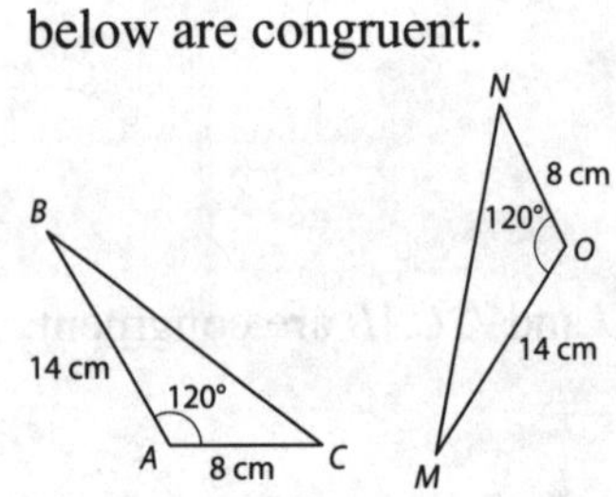

3. Explain why the two triangles below are congruent.

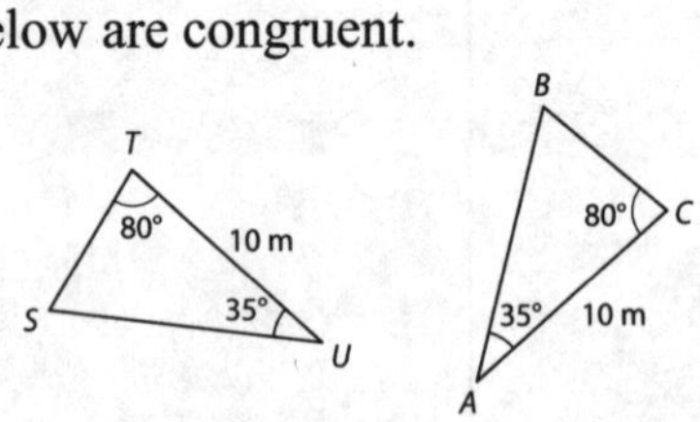

4. Determine whether the two triangles below are congruent.

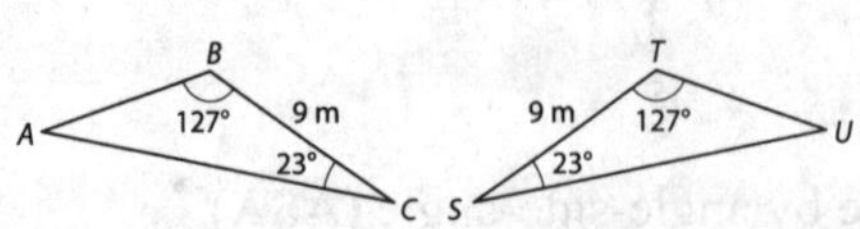

Name: ______________________________ Date: ______________

Instructor: ______________________________ Section: ______________

Notebook 9.8
Applications of Equations and Geometric Figures

OTHER NOTES

What are the four basic steps for problem solving?

1.

2.

3.

4.

EXAMPLE 1 Find the unknown width of the rectangular solid.

Understand the problem.
Create a plan.
Find the answer.
Check the answer.

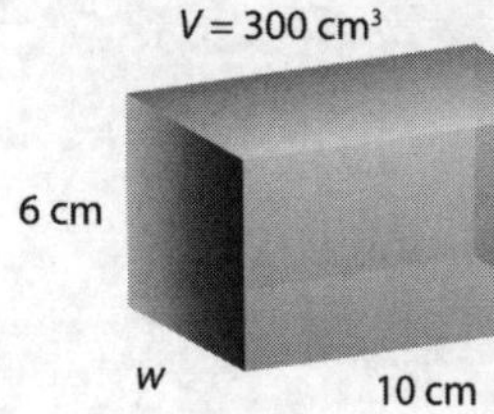

EXAMPLE 2 Find the length of the guy wire supporting the telephone pole.

Understand the problem.
Create a plan.
Find the answer.
Check the answer.

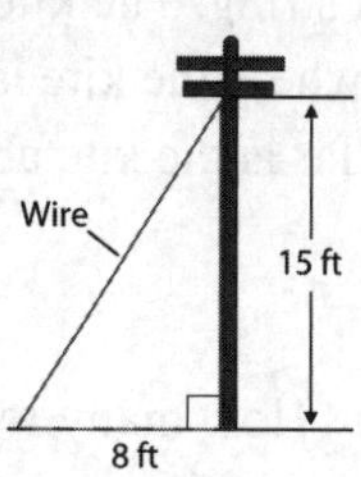

EXAMPLE 3 A manhole cover has a diameter of 3 ft. A strip of brass runs around the circumference of the cover. What is the length of this brass strip? Round to the nearest hundredth. (Use 3.14 for π.)

Understand the problem.
Create a plan.
Find the answer.
Check the answer.

EXAMPLE 4 A flagpole casts a shadow. At the same time, a small tree casts a shadow. Use the sketch to find the height of the flagpole.

Understand the problem.
Create a plan.
Find the answer.
Check the answer.

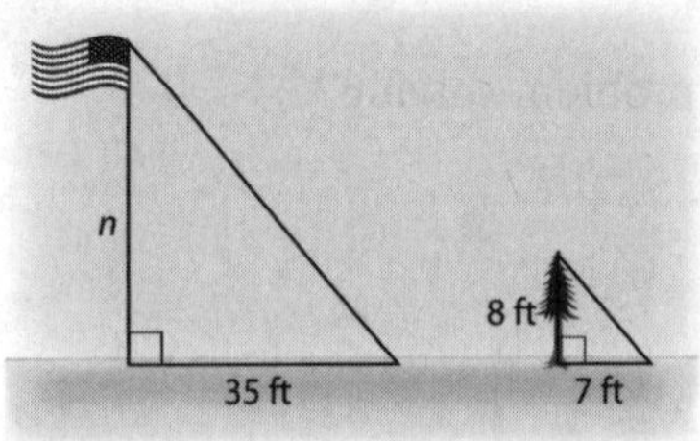

CONCEPT CHECK:
What formula is used to find the width of a rectangular solid?

GUIDED EXAMPLES:

1. Find the unknown length of the rectangular solid.

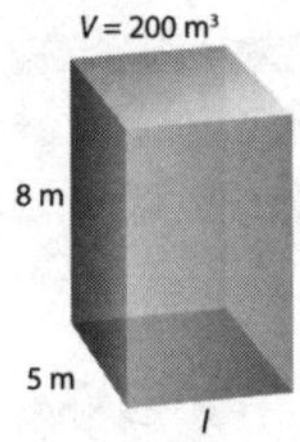

2. Barbara is flying her dragon kite on 32 yd of string. The kite is directly above the edge of a pond that is 30 yd from where the kite is tied to the ground. To the nearest tenth of a yard, how far is the kite above the pond?

3. Jimmy's truck has tires with a radius of 30 in. How many feet does his truck travel if the wheels make 9 revolutions? (Use 3.14 for π.)

4. A flagpole casts a shadow. At the same time, a small tree casts a shadow. Use the sketch to find the height n of the flagpole.

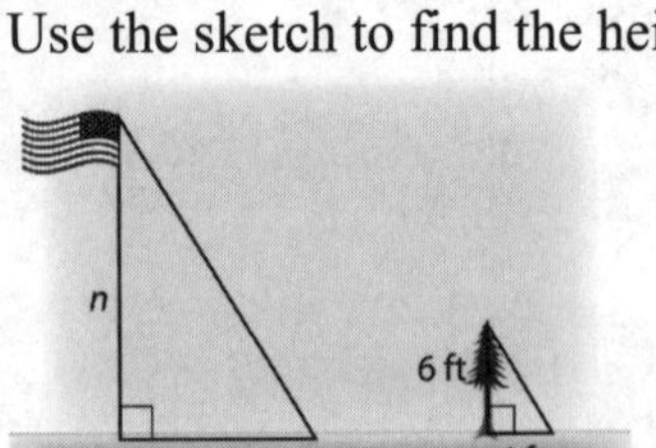

Name: ______________________________ Date: ________________

Instructor: ___________________________ Section: ______________

Notebook 10.1
Bar Graphs

OTHER NOTES

A bar graph can:

display _________
show ________over a period of ___________
show how often a particular data value ________
compare _______ types of _______
use __________________ to distinguish between
different ____________of data

EXAMPLE 1 Use the bar graph to answer the following questions.

1. How many total medals did the U.S. win in the 2008 Olympics?

2. Which type of medal did the U.S. win most often?

3. How many more silver medals were won than bronze medals?

EXAMPLE 2 Use the bar graph to answer the following questions.

1. How many Country Music Association Awards did Loretta Lynn win?

2. In which category did she win the fewest awards?

3. Write the ratio of the number of Entertainer of the Year awards won to the total number of awards won.

EXAMPLE 3 Use the double bar graph to answer the following questions.

1. In which game did the Phillies score the most runs?

2. In which game(s) did the Yankees score more runs than the Phillies?

3. In which game(s) did the largest difference between the numbers of runs scored by the two teams occur?

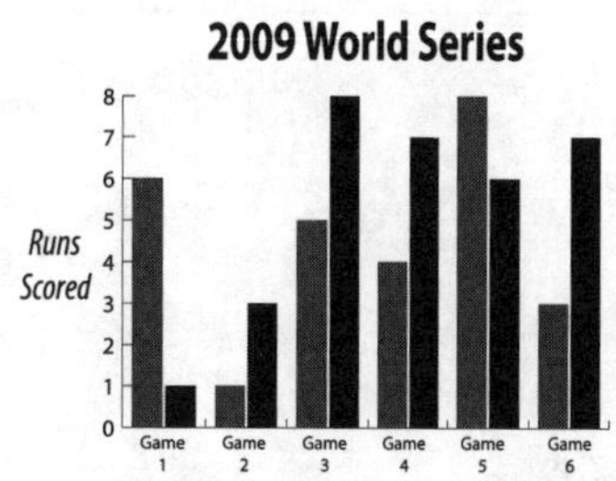

A histogram is a special type of ___________________.
In a histogram, the width of the bar represents an _____________,
or ____________, of numbers.
How does a bar graph differ from a histogram?

EXAMPLE 4 Use the histogram to answer the following questions.

a. What grade was earned most often?
How many times?

b. How many students were in the class?

c. How many students earned an F for the course?

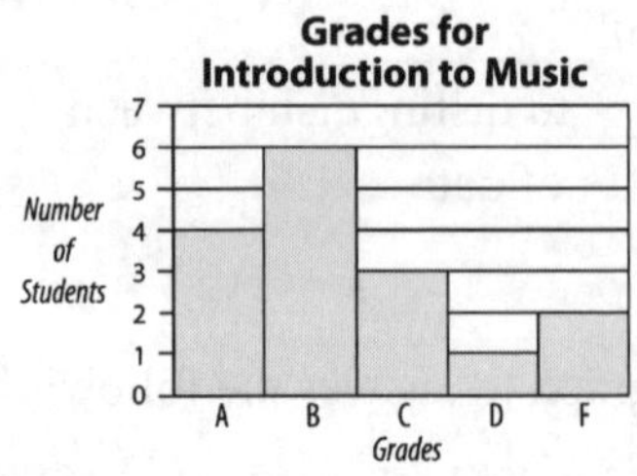

What kind of graph can be used to help organize data before creating a histogram?

EXAMPLE 5 In an Intro to Music class, letter grades are distributed based on the scale in the first column of Table 1. Complete the frequency distribution table for an Intro to Music class with scores of 78, 69, 82, 95, 92, 80, 47, 89, 81, and 99.
Then use the table to create a histogram displaying the same data.

TABLE 1 *Intro to Music Grade Distribution*		
Grade Intervals	Tally	Frequency
90–100		
80–89		
70–79		
60–69		
Below 60		

CONCEPT CHECK:
Why should a bar graph be used?

Name: ______________________________ Date: ______________

Instructor: ___________________________ Section: ____________

GUIDED EXAMPLES:

1. The bar graph shows the numbers of active coal mines in 2007 for several states.

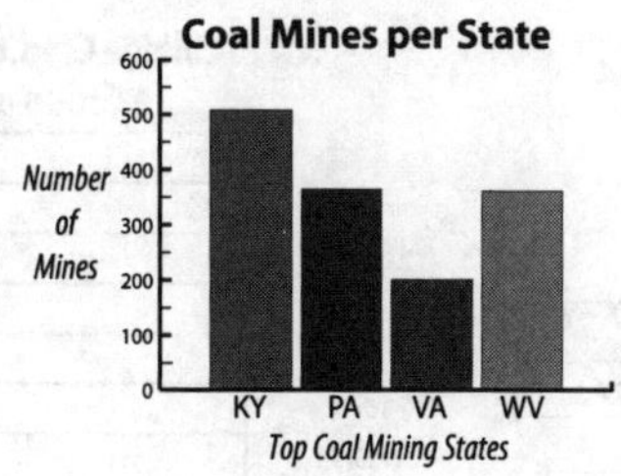

   a. Which state had the most mines? How many mines did it have?

   b. Which state had the fewest mines? How many mines did it have?

   c. Which states had the same number of mines? How many mines did each have?

2. The bar graph to the right shows the numbers of medals won by U.S. men and women and women at the World Figure Skating Championships.

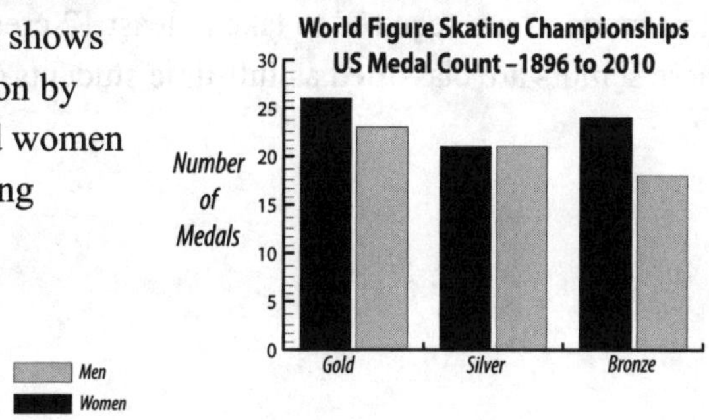

   a. Who won more gold medals—men or women? How many more gold medals were won?

   b. Which type of medal was won the same number of times by both men and women? How many times was it won?

   c. Which type of medal was won by the U.S. men the fewest times? How many times was it won?

3. The following frequency distribution table and histogram show the number of college credit hours taken by a group of 20 seniors.

College Credit Hours Taken by a Group of 20 Seniors		
Credit Hour Intervals	Tally	Frequency
0–3	/	1
4–7	///	3
8–11	//	2
12–15	~~////~~ ////	9
16 or more	~~////~~ /	6

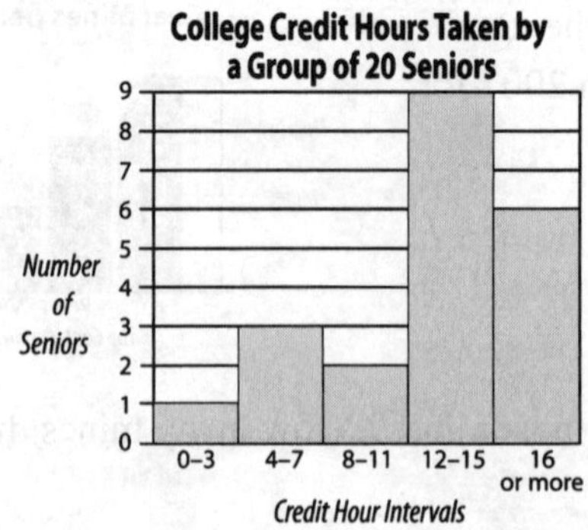

a. How many more seniors were taking 12–15 credit hours than were taking 4–7 credit hours?

b. How many seniors are taking 16 or more credit hours?

c. At most colleges, full-time students are required to take at least 12 credit hours. How many of these seniors are classified as full-time students?

Name: ______________________ Date: ____________

Instructor: ______________________ Section: ____________

Notebook 10.2
Line Graphs

A line graph can:

display ________

show ________ over a period of ____________

show how often a particular data value ____________

compare ________ types of ________

use data points that are connected with ____________

EXAMPLE 1 Use the line graph to answer the following questions.

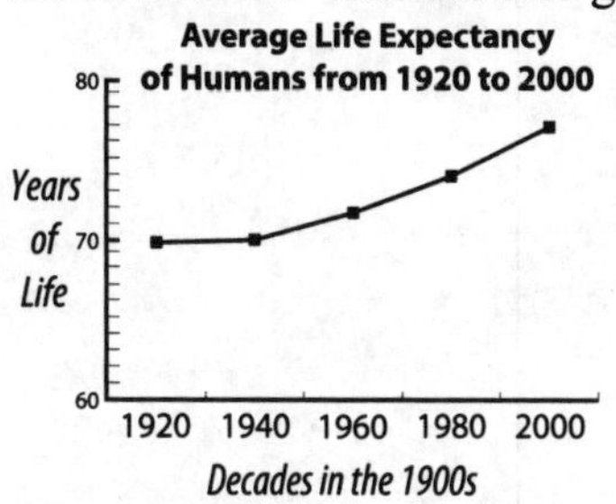

1. Has the average life expectancy of humans increased or decreased since 1920??

2. During which 20-year interval did the average life expectancy increase the most?

EXAMPLE 2 Use the line graph to answer the following questions.

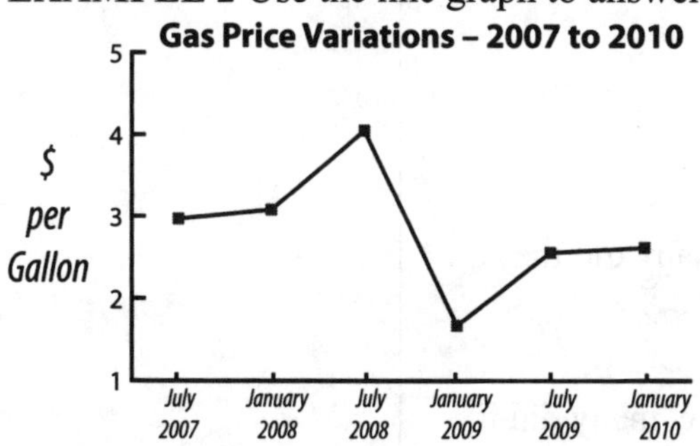

1. During which interval(s), or period(s) of time, did the largest increase/decrease in gas prices occur?

2. During the first half of 2009, did gas prices increase or decrease?

3. During which month and year does the graph show the highest gas price?

What kind of graph is used to show more than one set of data on a graph?

OTHER NOTES

EXAMPLE 3 Use the double line graph to answer the following questions.

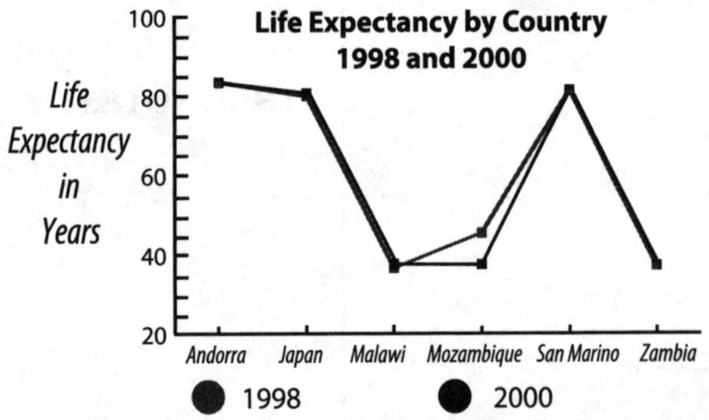

1. Which countries had the same life expectancy for 2000 and 1998?

2. What was the approximate life expectancy for Japan in 2000?

3. In which year was Mozambique's life expectancy lower, 2000 or 1998? How much lower was the life expectancy?

CONCEPT CHECK:

What is the difference between a bar graph and a line graph.

GUIDED EXAMPLES:

1.

a. In which year did the U.S. win the most medals? How many did they win?

b. In which year did the U.S. medal count decrease? By how many did it decrease?

c. In which year did the U.S. medal count increase the most over the previous Olympics? By how many?

d. In which years were the U.S. medal counts the same?

Name: ______________________________ Date: ______________

Instructor: ____________________________ Section: ____________

2.

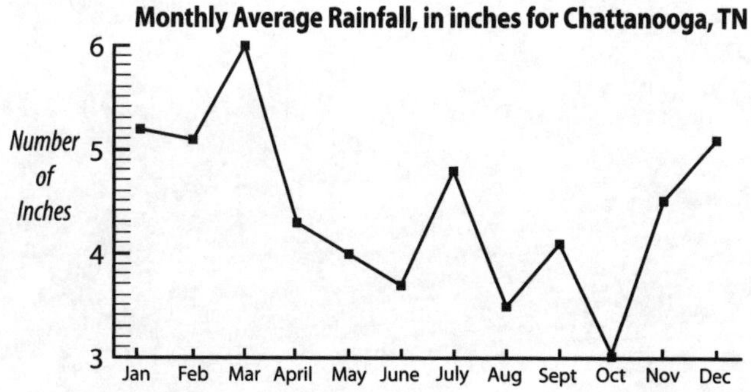

a. Which month had the highest average rainfall? What was the amount of rainfall?

b. Which month had the lowest average rainfall? What was the amount?

c. Between which two consecutive months did the greatest increase in rainfall occur? What was the difference in these two amounts?

3.

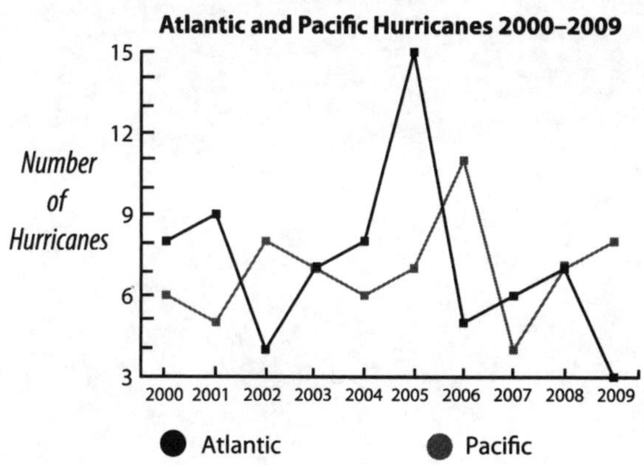

a. During which year did the most hurricanes occur in the Atlantic? How many hurricanes occurred in the Atlantic in that year?

b. During which year did the fewest hurricanes occur in the Pacific? How many hurricanes occurred in the Pacific in that year?

c. During which year(s) did more hurricanes occur in the Pacific than in the Atlantic?

d. During which year(s) did the Atlantic and the Pacific have the same number of hurricanes?

Name: ______________________________ Date: ________________

Instructor: ______________________________ Section: ______________

Notebook 10.3
Circle Graphs

OTHER NOTES

A circle graph can:
be used to show how ______________________
be drawn using actual __________ or __________

Notes:
In circle graphs, the circle represents the ______ or ______ amount.
All sections must combine to give the ____________ or __________.

EXAMPLE 1 Nikeshia received a scholarship at a local community college. The circle graph shows how she plans to use the money.

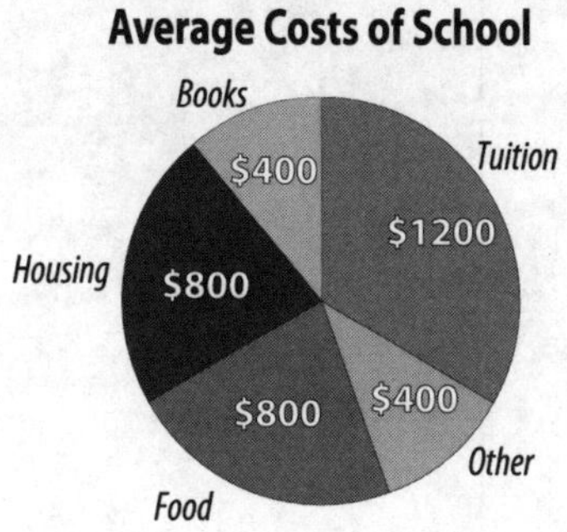

1. What does Nikeshia plan to spend the most money on?

2. What total amount does Nikeshia plan to spend on food and housing?

3. How much more does she plan to spend on tuition than on housing?

4. What is the total amount, in dollars, of Nikeshia's scholarship?

EXAMPLE 2 A Wildlife Management major at Cleveland State University needs 60 total credit hours to graduate with an Associate's Degree. The circle graph shows the number of credit hours needed in each category.

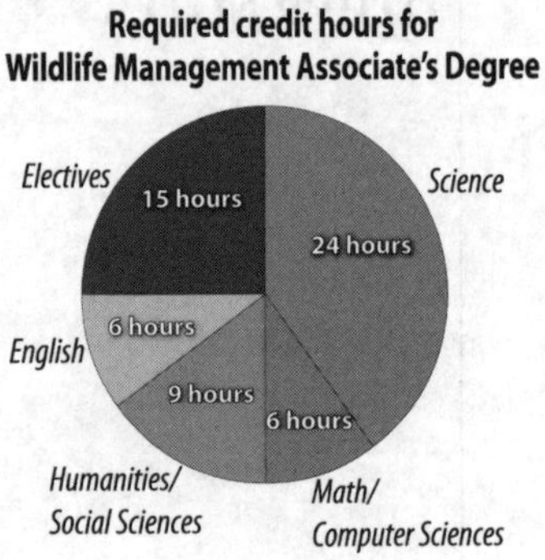

1. How many credit hours of electives are required for the degree?

2. What is the ratio of credit hours in Math/Computer Sciences to credit hours in Science?

3. How many credit hours are needed in English and Humanities/Social Sciences?

EXAMPLES 3 The circle graph shows how a student's final grade is calculated according to one instructor's syllabus.

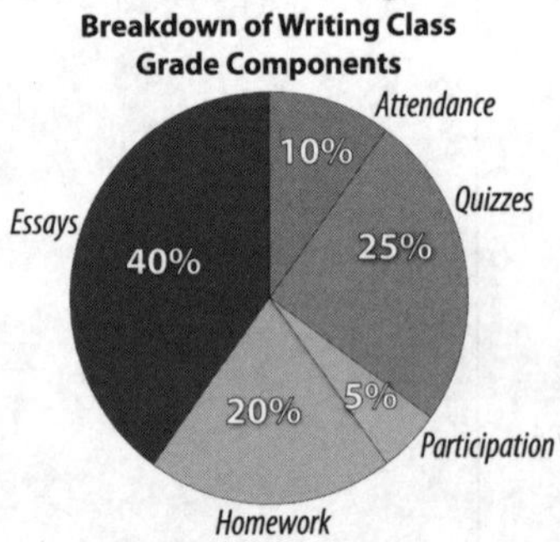

1. Which assignment type counts the most toward the student's final grade?

2. Assuming a student can earn a total of 2000 possible points in the class, how many points can be earned for attendance?

3. Assuming a student can earn a total of 1500 possible points, how many points can be earned for attendance?

CONCEPT CHECK:

True/ False A circular graph can be used with more than one set of data.

Name: ______________________________ Date: ________________

Instructor: ___________________________ Section: ______________

GUIDED EXAMPLES:

1. The circle graph shows how 70 people rated an album on Amazon.com from 1 (lowest) to 5 (highest).

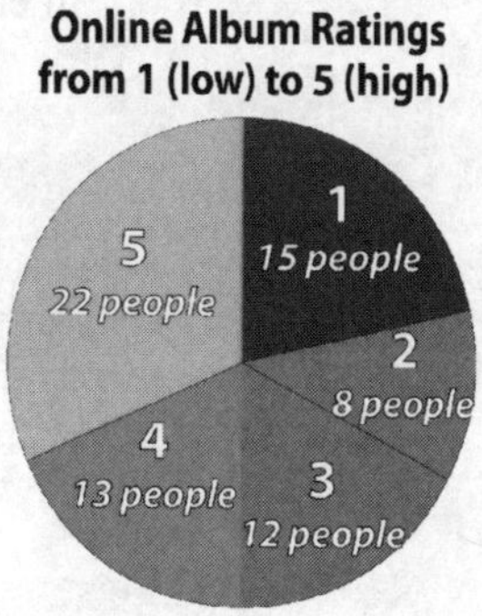

   a. What was the most popular rating?
   b. How many people rated the album as a 4?
   c. What is the ratio of the number of 2-ratings to the number of 3-ratings?
   d. How many people rated the album as a 1 or a 2?

2. The circle graph shows how Marcie spent her day.

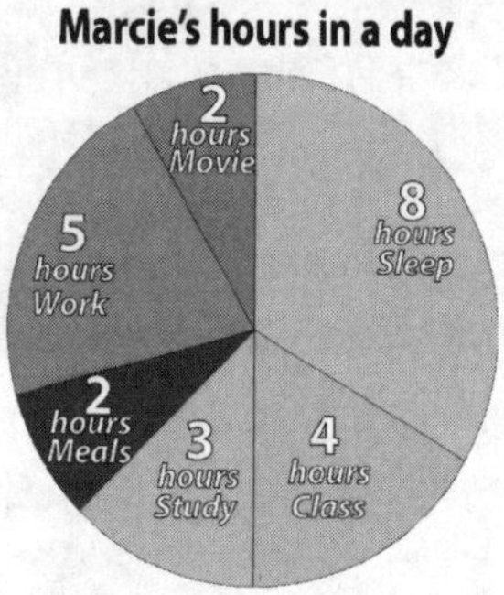

   a. How many hours did Marcie sleep?
   b. How did Marcie spend the greatest number of waking hours?
   c. How many hours did she spend on school, either in class or studying for class?
   d. How many more hours did she spend working than studying?

3. The circle graph shows U.S. mineral imports by country in 1993.

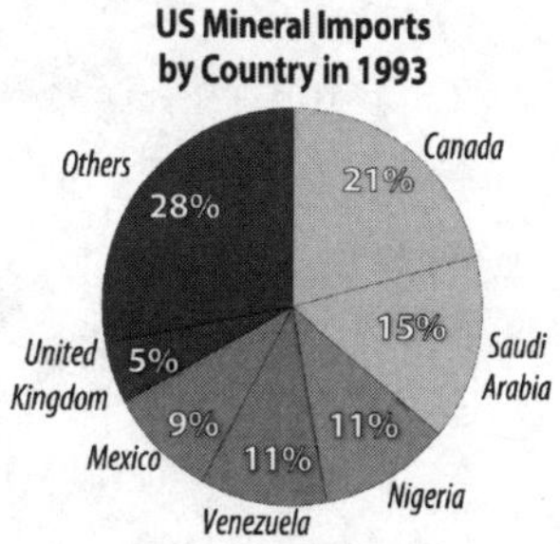

   a. What percent of minerals did the U.S. import from Mexico?
   b. What was the percent difference in the amount of mineral imports from Saudi Arabia and the amount of mineral imports from the United Kingdom?
   c. From which two countries did the U.S. import the same percent of minerals?

Name: ______________________ Date: ______________

Instructor: ______________________ Section: ______________

Notebook 10.4
Mean

OTHER NOTES

What is statistics?

__________ is information about a group or topic, often consisting of a set of _________.

What are three measurements commonly used to study the behavior of a set of numbers?

1.
2.
3.

These three measures are often referred to as ___________________ ___________________________. Why??

The mean is the __________ of the data.
The median is the ___________________ of the data.
The mode is the number(s) occurring ____________________ in the data.

Define mean.
The mean is the ____________ of the values in the list ______________ the number of _______________.

Mean = _________________________________

How do you calculate the mean? Write the steps.

1.

2.

EXAMPLE 1 Calculate the mean of the following list of numbers.

3, 5, 7, 2, 11, 6, 8, 2, 10

Sum =

Number of values =

EXAMPLE 2 Calculate the mean of the following list of test scores.

93, 86, 95, 98, 82

Sum =

Number of values =

EXAMPLE 3 Eli saves a portion of his allowance for 6 weeks. He saves $14 the first week, $12 the second week, $6 the third week, and $4, $10, and $2 for the remaining weeks. On average, how much does he save each week?

Understand the problem.

Create a plan.

Find the answer.

Check the answer.

CONCEPT CHECK:

Explain how to find the mean of 10, 15, and 20.

GUIDED EXAMPLES:

Calculate the mean of each list of numbers.

1. 7, 3, 11, 9, 14, 16, 20

2. 15, 21, 17, 22, 19, 25, 10, 22, 9, 16

3. Lynn went shopping and purchased items for $14.98, $7.60, $19.99, 11.95, and $8.53. Find the average cost of her purchases.

Name: ______________________________ Date: ______________

Instructor: ___________________________ Section: ______________

Notebook 10.5
Median

OTHER NOTES

______________ is the study and use of data.
__________ is information about a group or topic, often consisting of a set of
_________.

The _________, or average, is one statistical measure we can use when working with data.

The ____________, or middle number, is another.

What is the median?

What if there are two middle numbers?

Write the steps for finding the median of a list of numbers.
1.

2.

3.

Note: The numbers must be written in order from ___________ to _____________ before finding the median.

EXAMPLE 1 Calculate the median of the following list of numbers.

3, 5, 7, 2, 6, 8, 2, 10, 11
Ordered list:

Odd/even?

Middle number:

EXAMPLE 2 Calculate the mean of the following list of test scores.
44, 32, 31, 56, 77, 65
Ordered list:

Odd/even?

Average of two middle numbers:

EXAMPLE 3 A poll of nine students shows that the number of Facebook friends they have are 395, 486, 166, 430, 172, 159, 723, 582, and 319. Find the median number of Facebook friends for these nine students.

Understand the problem.

Create a plan.

Find the answer.

CONCEPT CHECK:

What is the median?

GUIDED EXAMPLES:

Find the median of each list of numbers.

1. 2, 5, 3, 7, 9, 5, 1, 3, 6, 9, 5

2. 11, 15, 13, 19, 18, 16, 11, 14

3. Seven friends worked part time for a week to earn money for a camping trip. The amounts they earned were \$66.00, \$48.75, \$86.90, \$41.00, \$46.50, \$87.50, and \$64.75. Find the median of the amounts earned.

Name: ______________________________ Date: ______________

Instructor: ____________________________ Section: _____________

Notebook 10.6
Mode

OTHER NOTES

Statistics is the study and use of ______.
Data is information about a _____ or topic, often consisting of a set of _________.

What is the mode?

How many modes can there be?

Notes:
Is it necessary to write the numbers in order before finding the mode?
Is a number that occurs more than once always a mode?

EXAMPLE 1 Find the mode(s) of the following list of numbers.

3, 5, 7, 2, 11, 6, 8, 2, 10

Ordered list:

Mode(s):

EXAMPLE 2 Find the mode(s) of the following list of numbers.

44, 32, 31, 56, 77, 65 65, 44, 65

Ordered list:

Number(s) occurring more often:

Mode(s):

EXAMPLE 3 Find the mode(s) of the following list of numbers.

325, 612, 367, 692, 451

Ordered list:

Number(s) occurring more often:

Mode(s):

EXAMPLE 4 Find the mode(s) of the following list of numbers.
2.1, 5.3, 6.2, 6.2, 7.5, 8.1, 8.1, 9.0, 11.4

Ordered list:

Number(s) occurring more often:

Number(s) occurring most often:

EXAMPLE 5 According to the National Weather Service, the high temperatures (in degrees Fahrenheit) for the first two weeks of June 2010 were recorded as 105, 106, 109, 112, 116, 119, 119, 115, 111, 107, 106, 103, 104, and 110. Find the mode(s) of the high temperatures for the time period.

Understand the problem.

Create a plan.

Find the answer.

Check the answer.

CONCEPT CHECK:
How many modes can there be for a given list of numbers?

GUIDED EXAMPLES:
Calculate the mode of each list of numbers.

1. 2, 5, 3, 7, 9, 5, 1, 3, 6, 9, 5

2. 25, 18, 30, 28, 42, 33, 51, 12

3. Over the past ten years, the numbers of hurricanes in the Atlantic Ocean were: 8, 9, 4, 7, 8, 15, 5, 6, 7, and 3. Find the mode(s) of the numbers of hurricanes during this 10 year period.

Name: ______________________________ Date: ________________

Instructor: ____________________________ Section: ______________

Notebook 10.7
Introduction to Probability

OTHER NOTES

What is the purpose of a tree diagram?
Draw a tree diagram for flipping a coin.

EXAMPLE 1 Draw a tree diagram showing the possible outcomes of rolling a single die.

What is probability?

What does a probability of 1 mean?
What does a probability of 0 mean?

How do you calculate probability?

EXAMPLES 2–4 Find the probability of the following events.
Tossing a coin and having it land on tails

Rolling a 4 on a single die

Rolling a 1 or a 3 on a single die

EXAMPLES 5–9 A bag contains 6 red marbles, 7 blue marbles, 2 black marbles, and 5 green marbles.
Find the probability of picking a red marble out of the bag.

Find the probability of picking a black marble out of the bag.

Find the probability of picking a blue or black marble.

Find the probability of picking a marble of any color.

Find the probability of picking a yellow marble.

EXAMPLES 10–12 A board game uses a colored spinner to determine where a player will move on his or her next turn.

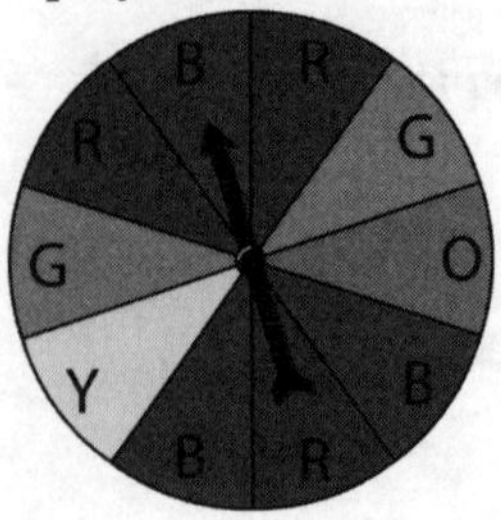

Find the probability of the spinner landing on red.

Find the probability of the spinner landing on green.

Find the probability of the spinner landing on a primary color. (red, yellow, or blue).

CONCEPT CHECK:

True/False A tree diagram showing the outcomes of a 6-sided die will have 7 branches.

GUIDED EXAMPLES:

1. Draw a tree diagram showing the number of outcomes of throwing an octahedral (8-sided) die with each side representing a number 1 through 8.
2. Find the probability of the following event.
   Tossing a coin and getting heads

A bag contains 1 red marble, 5 green marbles, 2 blue marbles, and 2 black marbles. Find the probability of the following events.

3. Picking a blue marble
4. Picking a blue or green marble

Name: ______________________________ Date: ________________

Instructor: ___________________________ Section: ______________

Notebook 11.1
Signed Numbers

OTHER NOTES

Recall: Whole numbers are used to represent "whole" quantities.
Fractions and decimals are used to represent "parts" of a whole.

Signed numbers are used to represent positive and negative numbers.
A positive number is a number that has a value _____ than 0.
A negative number is a number that has a value _____ than 0.

What sign is used to represent a positive number?
What sign is used to represent a negative number?

Note: You must use the "–" sign with a negative number to indicate that the number's value is ______ than 0.

Give examples of uses for negative numbers.

Recall the symbols used for comparing numbers.
greater than ....
less than ....

What is the purpose of a number line?

EXAMPLES 1–3 Which number is larger?

5 or 8

–3 or 2

–7 or –10

EXAMPLES 4 & 5 Insert a < or > in the blank to make each statement true.

4 _____ –4

–8 _____ –10

What are opposites?
Give an example.

The __________ of a number is written using a negative (or minus) sign.

The opposite of positive 5 is ....
The opposite of negative 10 is ....

What are the purposes of the minus sign?
1.

2.

3.

What is the procedure for finding the opposite?
What is the opposite of 0?

EXAMPLES 6 & 7 Write the following in symbols. Then simplify if necessary.

The opposite of 8

The opposite of –25

EXAMPLES 8 & 9 Find the opposite.

–16

–38

What are the two parts of a number?

What is the absolute value of a number?

EXAMPLES 10–12 Find the absolute value.

$|4|$

$|-13|$

$|9-7|$

EXAMPLES 13–16 Simplify.

$-|-12|$

$-|10.6|$

$-|0|$

$-|7-1|$

Note: $-(-9) = 9$, but $-|-9| = -9$.

CONCEPT CHECK:
Why is it impossible for an absolute value to be negative?

GUIDED EXAMPLES:

1. Insert a < or > in the blank to make the statement true.
   –11 ___ –6

2. Find the opposite and simplify.
   –19

3. Simplify.
   $-|11-7|$

Name: ______________________________ Date: ________________

Instructor: ____________________________ Section: ______________

Notebook 11.2
Adding Signed Numbers

OTHER NOTES

What is the chip method?
What kind of chip is used for positive numbers?

What is the number line method for adding positive numbers?

What kind of chip is used for negative numbers?

What is the number line method for adding negative numbers?

The absolute value of a number is its ________ from zero on a number line.
Absolute value is represented by two ____________ bars.
The absolute value of a number, like distance, can _________ be a negative number.

What is the procedure for adding two numbers with the same sign?

EXAMPLES 1–3 Add.

$-9 + (-11)$

$3.25 + 4.16$

$-\frac{3}{11} + \left(-\frac{4}{11}\right)$

What are opposites?

What is the rule for the sum of two opposites (additive inverses)?
How can chips be used to represent additive inverses?

What is the number line method for adding positive and negative numbers?

What is the procedure for adding two numbers with different signs?

EXAMPLES 4 & 5 Add.

$-18 + 7$

$6 + (-2.5)$

EXAMPLES 6 & 7 Add.

$3 + (-5) + (-2)$

$-\frac{5}{9} + \frac{1}{3}$

CONCEPT CHECK:

If adding three or more numbers with different signs, can you first add all of the positive numbers together, then add all of the negative numbers, then apply the procedure for adding numbers with different signs?

GUIDED EXAMPLES:

1. Use the chip method to add.
   $-3 + (-7)$

2. Use the number line method to add.
   $-4 + (-6)$

3. Use the chip method to add.
   $8 + (-5)$

4. Use the chip method to add.
   $7 + (-11)$

5. Add.
   $-1.6 + (-2.1)$

6. Add.
   $-3.5 + 2.6$

7. Add.
   $-11 + 19$

8. Add.
   $7 + (-4.25)$

Name: ______________________________ Date: ________________

Instructor: ____________________________ Section: ______________

Notebook 11.3
Subtracting Signed Numbers

OTHER NOTES

To add two numbers with the same sign, ____ the absolute values (or numerical parts). The sign of the sum is the same as the sign of the original numbers.

To add two numbers with different signs, ________ the absolute values (or numerical parts). The sign of the answer is the same as the sign of the number with the larger absolute value (the number that is farthest from 0 on a number line).

To subtract, change the subtraction (minus) sign to an addition (plus) sign and then change the sign of the number being subtracted to its opposite.

What is "LEAVE, CHANGE, CHANGE?"

What is the procedure for subtracting signed numbers?

EXAMPLES 1–4 Add or subtract.

$6 - 4$

$6 + (-4)$

$5 + (-9)$

$5 - 9$

EXAMPLES 5 & 6 Subtract.

$-7 - 3$

$4 - (-8)$

Note: When subtracting a negative number, the rewrite will result in adding a positive number.

EXAMPLES 7–9 Subtract.

$2 - 5$

$9.4 - (-2.1)$

$-\frac{3}{4} + \left(-\frac{5}{4}\right)$

EXAMPLES 10–13 Subtract.

$9 - 7$

$9 - (-7)$

$-9 - 7$

$-9 - (-7)$

EXAMPLES 14–16 Subtract.

$-14 - (-14)$

$-5.6 - 9.3$

$6 - (-8) - 4$

EXAMPLE 17 Today at 4:00 p.m., the temperature outside was –4ºF. Three hours later, the temperature had dropped 9 degrees. What was the temperature (in degrees Fahrenheit) at 7:00 p.m.?

Understand the problem.

Create a plan.

Find the answer.

Check the answer.

CONCEPT CHECK:

How does the concept of absolute value help to explain the statement "Subtracting a negative number is the same as adding a positive number?" Use 2 – (–6) as an example.

GUIDED EXAMPLES:

1. Subtract. $-4 - 7$

2. Subtract. $12 - (-8)$

3. Subtract. $-14 - (-20)$

4. Subtract. $-\frac{3}{5} - \frac{2}{5}$

5. Today at 4:00 p.m., the temperature outside was –8ºF. Three hours later, the temperature had dropped 10 degrees. What was the temperature (in degrees Fahrenheit) at 7:00 p.m.?

Name: ______________________________ Date: ____________

Instructor: ____________________________ Section: ____________

Notebook 11.4
Multiplying and Dividing Signed Numbers

OTHER NOTES

What method can you use to remember the rules for multiplying and dividing signed numbers?

What is the procedure for multiplying signed numbers?

EXAMPLES 1–3 Multiply.

$(-9)(-11)$

$8(-7)$

$-\frac{3}{4}\cdot\left(-\frac{4}{5}\right)$

EXAMPLES 4–7 Multiply.

$(-1)(-2)$

$(-1)(-2)(-3)$

$(-1)(-2)(-3)(-4)$

$(-1)(2)(-3)(4)$

Notes:

When multiplying an _______ number of negative factors, the product is positive.

When multiplying an odd number of negative factors, the product is ___________.

What is used as a shortcut for repeated multiplication?

In the expression 2^3, 2 is the _____ and 3 is the _________.

EXAMPLES 8–10 Identify the base and exponent, then evaluate.

4^2	3^3
$(-4)^2$	$(-3)^3$
-4^2	-3^3

When a negative base is raised to an _____ power, the result is always positive.

When a negative base is raised to an odd power, the result is always ________.

Write the rule for negative numbers and exponents using symbols.

What is the procedure for dividing signed numbers?

To divide two numbers with the same sign, divide the numerical parts (absolute values). The quotient is ________ .

To divide two numbers with ________ signs, divide the numerical parts (absolute values). The quotient is negative.

EXAMPLES 11 & 12 Divide.

$-18 \div (-6)$

$-24 \div 2$

Is the statement true or false? Why

$$\frac{-2}{3} = \frac{2}{-3} = -\frac{2}{3}$$

EXAMPLES 13–15 Divide.

$\frac{-45}{-9}$

$14 \div (-3.5)$

$\frac{-32}{6}$

CONCEPT CHECK:

Will the result of $(-4)^3 \div (-8)^2$ be positive or negative?

GUIDED EXAMPLES:

1. Multiply. $(-5)(-12)$
2. Multiply. $11(-13)$
3. Multiply. $6(-4)(-2)(5)(-3)$
4. Evaluate. $-(-3)^4$
5. Divide. $18 \div 3$
6. Divide. $(4.5) \div (-0.5)$
7. Divide. $(-12) \div (-4)$

Name: ______________________________ Date: ______________

Instructor: ___________________________ Section: ____________

Notebook 11.5

Order of Operations and Signed Numbers

OTHER NOTES

What is the procedure for the order of operations?

1.
2.
3.
4.

Note: [ ], { }, and fraction bars can also act like parentheses.

What is the purpose of Please Excuse My Dear Aunt Sally?

What is the procedure for adding signed numbers?

To add two numbers with the _____ sign, add the absolute values (or numerical parts). The sign of the sum is the sign of the original numbers.

To add two numbers with different signs, ________ the absolute values (or numerical parts). The sign of the answer is the sign of the number with the larger absolute value (the number that is farthest from 0 on the number line.

What is the procedure for subtracting signed numbers?

To subtract two signed numbers, _____ the first number alone, ______ the minus sign to a plus sign, and _____ the sign of the number being subtracted to its opposite. Then follow the rules for addition.

What is the procedure for multiplying and dividing signed numbers?

To multiply (or divide) two numbers with the _____ sign, multiply (or divide) the numerical parts (absolute values). The product (or quotient) is positive.

To multiply (or divide) two numbers with different signs, multiply (or divide) the numerical parts (absolute values). The product (or quotient) is _________ .

EXAMPLE 1 Simplify.

$(-5)^2 + 3^2$

Parentheses

Exponents

Multiply and Divide

Add and Subtract

EXAMPLE 2 Simplify.

$$(-6+4)^3 \div (-2)$$

Parentheses
Exponents
Multiply and Divide
Add and Subtract

EXAMPLE 3 Simplify.

$$(3-6)^2 - 2(-10)$$

Parentheses
Exponents
Multiply and Divide
Add and Subtract

EXAMPLE 4 Simplify.

$$\frac{(-4)+6-3}{4^2+(2)(-3)}$$

CONCEPT CHECK:
To simplify $(-9+6)-2(-5)$, which operation will you perform first?

GUIDED EXAMPLES:

1. Simplify. $3+(-4)2-(-10)$
2. Simplify. $(-77)\div\left(8-(-3)\right)-9$
3. Simplify. $-5\cdot(12-15)^2-30$
4. Simplify. $\dfrac{-8-3(2)}{(-4)^2-(-4)}$

Name: ______________________ Date: ______________

Instructor: ______________________ Section: ______________

Notebook 11.6
Applications of Operations with Signed Numbers

OTHER NOTES

What is the procedure for adding two numbers with the same sign?

What is the procedure for adding two numbers with different signs?

EXAMPLES 1 & 2 Add the following.

$-2 + (-4)$

$1 + (-8)$

What is the procedure for subtracting signed numbers?

Leave, change, change

EXAMPLES 3 & 4 Subtract the following.

$-2 - (-6)$

$-3 - 5$

EXAMPLE 5 Emmanuel will be driving several days to get home for a reunion. At one point on the first day of his trip, his elevation was 750 feet below sea level. The next day his elevation dropped another 450 feet. What was Emmanuel's elevation at the end of the second day?

Understand the problem

Create a plan

Find the answer

Check the answer

EXAMPLE 6 At the end of June, the rainfall for the year, so far, was 6.5 inches below normal. During the second half of the year, it rained 11.25 inches above normal. What amount of rainfall fell above or below normal for the entire year?

Understand the problem

Create a plan

Find the answer

Check the answer

What is the procedure for multiplying signed numbers?

EXAMPLES 7–9 Multiply the following.

$(-6.1)(-2)$

$-\frac{2}{5}\cdot\frac{3}{4}$

$(-4)(5)(-3)$

What is the procedure for dividing two numbers with the same sign?

What is the procedure for dividing two numbers with different signs?

EXAMPLES 10 & 11 Divide the following.

$-64.8 \div (-8)$

$\frac{8}{3} \div \left(-\frac{5}{9}\right)$

Name: ______________________________ Date: ________________

Instructor: ___________________________ Section: ______________

EXAMPLE 12 An airplane flying to Knoxville was descending at a rate of 2750 feet per minute. If the plan continues at this rate, what is the overall change in altitude after 8 minutes?

Understand the problem

Create a plan

Find the answer

Check the answer

EXAMPLE 13 Cade is snowboarding down a mountain at a rate of 63 meters every 7 seconds. What is the average change of elevation per second? That is, how much does Cade's elevation change every second, on average?

Understand the problem

Create a plan

Find the answer

Check the answer

CONCEPT CHECK:
What are the four steps to solving an application problem?

GUIDED EXAMPLES:

1. Whiteface Mountain has a vertical drop of 3430 feet. Assuming an Olympic skier can ski down the mountain in 28 seconds, what is the average change in the vertical drop per second?

2. On January 1, Callie weighed 182 pounds. One month later she weighed 177. Find the overall change in Callie's weight if she continued to lose at this rate for 4 months.

3. At 12:30 A.M., a submarine was 614 meters below the surface. Over the next three hours, the submarine continued to descend 127 meters deeper. How deep is the submarine at the end of the three hours?

4. In a certain city, the average temperature for January is –9.8° Fahrenheit. The average temperature for February is –2.6° Fahrenheit. What is the increase in the average temperature from January to February?

5. At the end of each work day, a cashier must balance his or her register and record any discrepancies. A new cashier has daily discrepancies of $4.14, –$3.27, –$2.99, $0.19, and –$1.12. What is the average daily discrepancy for this cashier for the week?

6. A hot air balloon is descending at a rate of 130 feet per minute. If the balloon continues at this rate, what is the overall change in altitude after 12 minutes?

7. Barry thought he had $56.12 in his checking account, but he forgot to write down a withdrawal of $50.00. When he then used his debit card to purchase food for $8.75, he was also charged an insufficient funds fee of $15.00. What is the actual balance in Barry's checking account after the food purchase and fee?

8. Austin, the Wildcat's quarterback, threw a pass for a 15-yard gain, but was then sacked for a 17-yard loss. During the next couple of plays, he ran for a gain of 6 yards and then again for a 1-yard gain. What is the total gain/loss of these four plays?

Name: ______________________________ Date: ______________

Instructor: ___________________________ Section: ____________

Notebook 12.1
Variables and Algebraic Expressions

OTHER NOTES

What is a variable?

Notes:

Why is the letter O not used as a variable?
When writing variables, be careful to make sure that they do not look like __________ .

Name some examples of formulas.

In algebra, which is an area of mathematics in which we work with unknown values in order to solve problems, we use variables to ____________ the unknown values.

What is a term?
Give an example of a term.

Every term has two parts: a __________ (numerical coefficient or simply coefficient) and a variable or variables.

EXAMPLES 1–3 Identify the coefficient and the variable(s).

$-4x$

$3t$

$-8abc$

What is an expression?
Give some examples.

What is an algebraic expression?
Give some examples.

Note: A _______ can also be an algebraic expression. It is simply an algebraic expression that contains one term. We will be using only algebraic expressions from this point forward in the course.

EXAMPLES 4–6 Decide if each of the following is a term, an algebraic expression, or both.

$57g$

$4t - 8v$

$x + y + 2$

To "evaluate an algebraic expression" means to ______ the value of the expression when ______ a specific value for each variable.

What is the procedure for evaluating algebraic expressions?

1.

2.

EXAMPLES 7 & 8 Evaluate the algebraic expression.

$3n$ for $n = 7$

$x + y - 5$ for $x = 2.5$ and $y = 9.1$

EXAMPLES 9–11 Evaluate each algebraic expression.

$3m - 5n$ for $m = -4$ and $n = 2$

$\frac{8p}{q}$ for $p = 6$ and $q = -3$

$-5(x + y)$ for $x = 4$ and $y = -9$

We can translate from words to symbols using ________ as well as numbers.

EXAMPLES 12–14 Translate from words to symbols. Use n as the unknown number (the variable).

the difference of 4 and a number

twice the sum of 3 and a number

the product of –6 and a number

CONCEPT CHECK:

How many terms in the expression $\frac{1}{2} + 2p$? Is it an algebraic expression?

GUIDED EXAMPLES:

1. For the given term, identify the coefficient and the variable(s).

   $-8uvw$

2. Evaluate the following.

   $\frac{m+n}{p-q}$ for $m = 5$, $n = 3$, $p = 6$, and $q = 2$

3. Translate from words to symbols. Use n as the unknown number (the variable).

   Four times the sum of a number and eleven

Name: ______________________________ Date: ________________

Instructor: ____________________________ Section: ____________

Notebook 12.2
Like Terms

OTHER NOTES

A ____________ is any number, variable, or product of numbers and/or variables.

Terms that differ only by the numerical coefficients are called ______ terms.

State the definition of like terms.
Give some examples.

Like terms can be __________ without variables too.

Are $2x$, $2y$, and $2z$ like terms? Why or why not?

EXAMPLES 1–3 Circle the like terms in each list.

$4a$, 9, $13a$

$-5st$, st, $7t$, $18st$, $9s$

2, $6s$, 33, $9y$, -1

Note: Only like terms can be added or subtracted.

When an expression is simplified by adding or subtracting like terms, this is often referred to as ________ like terms.

What is the procedure for combining like terms?

EXAMPLES 4 & 5 Combine like terms.

$13a - 4a + 6a$

$7m - 3m - 6m$

EXAMPLES 6 & 7 Simplify by combining like terms.

$6c + 3c - 13c + c$

$-9a - 5a + 2a$

EXAMPLES 8 & 9 Simplify (combine like terms).

$5x + 8 - 9x + 2$

$2 + (-3y) - 4 + 6y$

EXAMPLES 10 & 11 Simplify (combine like terms).

$4a - 5b + 6a + 7$

$\frac{3}{4}x + \frac{2}{5}y - \frac{1}{4}x + \frac{1}{5}y + 3$

CONCEPT CHECK:

Are y^2 and $2y$ like terms? Explain.

GUIDED EXAMPLES:

1. Combine like terms.

   $5x - 2x - 6x$

2. Combine like terms.

   $-6ab + 10ab + 3ab$

3. Simplify by combining like terms.

   $7y + 2x - 5y + 8x$

4. Simplify by combining like terms.

   $3s - 5 - 9s + 14$

5. Simplify by combining like terms.

   $\frac{2}{5}x + \frac{3}{4}y - \frac{3}{4}x + \frac{6}{7}$

Name: ______________________________ Date: ________________

Instructor: ____________________________ Section: ______________

Notebook 12.3
Distributing

OTHER NOTES

State the Distributive Property of Multiplication.

When multiplying a number by a sum or difference, multiply the number __________ the parentheses by each number __________ the parentheses.

This process is called __________ and is referred to as the Distributive Property.

When is the Distributive Property used?

Note: The ___________________ indicates that we perform the operation inside the parentheses first, but this is not always possible. When this occurs, we use the Distributive Property to rewrite the expression without ___________ .

EXAMPLE 1 Rewrite using the Distributive Property and simplify if possible.

$4(x+3)$

EXAMPLES 2–4 Multiply using the Distributive Property.

$7(x-5)$

$-2(a+9)$

$(y-3)(-4)$

The Distributive Property can be used to multiply:

when the terms inside….

when more than two terms ….

when the expression is written …

EXAMPLES 5 & 6 Multiply using the Distributive Property.

$(6x+9)5$

$10(3a-b+5c)$

EXAMPLES 7–9 Multiply using the Distributive Property.

$\frac{1}{2}(4x+8)$

$\frac{2}{3}(3x-12)$

$-\frac{3}{5}(x-4)$

CONCEPT CHECK:

Does the expression $4(xy)$ require the Distributive Property to simplify?

GUIDED EXAMPLES:

Multiply using the Distributive Property.

a. $8(x+5)$

b. $-4(2a+3b-c)$

c. $(x-2y+4z)6$

d. $-\frac{3}{4}(20t-32)$

Name: ______________________________ Date: ________________

Instructor: ______________________________ Section: ______________

Notebook 12.4
Simplifying Expressions

OTHER NOTES

An __________ includes one or more terms, with multiple terms separated by "+" or "–" signs.

An algebraic expression is an expression that contains one or more ______________ .

When an algebraic expression is written in _____________ form, none of the terms can be combined.

The algebraic expression $3x + 5x$ simplifies to ______ .

The algebraic expression $6x + 4$ is in __________ form.

If an algebraic expression contains parentheses and it is not possible to simplify what is inside the parentheses first, what property is applied to remove the parentheses and combine like terms?

EXAMPLES 1–3 Simplify by using the Distributive Property and then combining like terms.

$5 + 2(3x - 6)$

$-4(a + 6) - 2a$

$2(3x + 5y) + 4x$

EXAMPLES 4 & 5 Simplify. (Use the Distributive Property and then combine like terms.)

$8(4x + 5) + 2(3x - 1)$

$\frac{2}{3}(6x + 3) + \frac{1}{4}(12x - 8)$

EXAMPLES 6 & 7 Simplify.

$5x-(3x+11)$

$4(y+2)-3(6+5y)$

CONCEPT CHECK:
Why are parentheses removed before combining like terms when simplifying algebraic expressions?

GUIDED EXAMPLES:

1. Simplify (Use the Distributive Property and then combine like terms).

   $3(2y+4)+9$

2. Simplify.

   $4x+2(5-6x)+7$

3. Simplify.

   $7(4a+2)-5(3a-6)$

4. Simplify.

   $\left(-\frac{1}{2}\right)(8x+12y)+\left(\frac{3}{5}\right)(10x-25y)$

Name: ______________________________ Date: ______________

Instructor: ____________________________ Section: ______________

Notebook 12.5
Equations and Solutions

OTHER NOTES

An __________ is a collection of one or more terms, with multiple terms separated by "+" or "–" signs.

An algebraic expression is an expression that contains one or more ____________ .

What is an equation?

What is the difference between an expression and an equation?

Give an example of an expression and an example of an equation.

What symbol does an equation have that an expression does not have?

An expression may be ___________ , while an equation may be solved.

EXAMPLES 1–5 Identify each as an expression or an equation.

$3(x-6)$

$4x+3=7$

$2x+4y-8$

$7(x-4)+3=4x-2$

$x=8$

Expressions may be ____________ by combining like terms.

Equations may be _____________ by finding the value(s) of the variable(s) that make the equation true.

What is the definition of the solution of an equation?

Note: An equation may have one solution, more than one solution or no solution.

What is the procedure to determine if a given value is a solution?

EXAMPLES 6 &7 Determine whether the given value is a solution of the equation.

Determine if 1 is a solution of $2x-3=1$.

Determine if –1 is a solution of $x+5=2(x+3)$.

The equation $2x+5=9$ can also be written as $9=2x+5$.
These two equation may look different, but they have the same __________ .
They are called equivalent equations.

What are equivalent equations?

EXAMPLE 8 Determine if the following are equivalent equations.
$2x+5=9$, $2x=4$, and $x=2$.

Note: When trying to find the solutions of equations, we often use equivalent equation to make the equations easier to solve.

CONCEPT CHECK:
How can you determine if a given number is a solution to an equation?

GUIDED EXAMPLES:

1. Identify as an expression or equation: $4x-9=5$.

2. Determine if –2 is a solution of the equation. $3x+4=2$

3. Determine if 5 is a solution of the equation. $3(x-2)=x+4$

4. Determine if these equations are equivalent.
   $4x=12$, $2x+1=6$, $x=3$

Name: ______________________________ Date: ________________

Instructor: ___________________________ Section: _____________

Notebook 12.6
Solving Equations by Adding or Subtracting

OTHER NOTES

A ____________ of an equation is a number or numbers that, when substituted for the variable, makes the equation true.

How do you determine if a given value is a solution of an equation?

What is the procedure to solve an equation?

To solve an equation, reverse operations are often needed.

The reverse operation of addition is ….

The reverse operation of subtraction is ….

Addition and subtraction "undo" each other, which means that adding a number to, and subtracting the *same* number from, an expression results in ____________ .

What is the addition property of equality?

Write the addition property of equality using symbols.

Note: Since subtraction can be defined in terms of addition, this property works for subtraction as well as addition. This means that you can also subtract the same number from ______ sides of an equation without changing the solution.

EXAMPLE 1 Use the Addition Property of Equality to solve the equation $x - 3 = 8$.

What is the procedure for solving equations by adding or subtracting?

1.

2.

3.

4.

EXAMPLES 2 & 3 Solve each equation by adding or subtracting. Check your solution.

$x + 2 = -9$

$x - 5 = 3$

Always remember to keep the equation __________ . If you add a number to one side of the equation, you must add the same number to the other side.

If you subtract a number from one side of the equation, you must subtract the ________ number from the other side.

EXAMPLES 4 & 5 Solve each equation by adding or subtracting. Check your solutions.

$x - 9 = -5$

$2 = x + (-7)$

EXAMPLES 6 & 7 Solve each equation by adding or subtracting. Check your solutions.

$-2.5 = x + 1.3$

$x - \frac{3}{7} = -\frac{2}{7}$

CONCEPT CHECK:

Explain how the addition property of equality is used to solve $x + 2 = 1$. How does this differ from how it would be used to solve $x - 2 = 1$?

GUIDED EXAMPLES:

1. Solve by adding or subtracting. Check your solution. $x - 6 = 7$

2. Solve by adding or subtracting. Check your solution. $x + 3 = -8$

3. Solve by adding or subtracting. Check your solution. $x - \frac{3}{4} = -\frac{5}{6}$

4. Solve by adding or subtracting. Check your solution. $x + 2.4 = 1.9$

Name: ______________________________ Date: ________________

Instructor: ___________________________ Section: ______________

Notebook 12.7
Solving Equations by Multiplying or Dividing

OTHER NOTES

Recall Addition and subtraction are reverse operations.
Likewise, multiplication and division are reverse operations.

What is the multiplication property of equality?

Note: Since division can be performed by multiplying by the ____________ , this property works for division as well as for multiplication.

EXAMPLE 1 Use the Multiplication Property of Equality to solve the equation $3x = 18$.

What is the procedure for solving equations by multiplying or dividing?

1.

2.

3.

EXAMPLE 2 Solve by multiplying or dividing. Check your solution. $4x = -8$

EXAMPLES 3–5 Solve by multiplying or dividing. Check your solutions.

$-6x = -12$

$\frac{x}{5} = 15$

$\frac{x}{-4} = 3$

EXAMPLES 6 & 7 Solve by multiplying or dividing. Check your solutions.

$0.4x = -7$

$-\frac{x}{6} = -4$

EXAMPLE 8 Solve by multiplying or dividing. Check your solutions.

$\frac{3x}{4} = -\frac{2}{3}$

Note: When the coefficient of x is a fraction, multiply both sides of the equation by the reciprocal of the coefficient. This makes the coefficient of x become 1.

CONCEPT CHECK:

Why is an equation solved by division instead of multiplication?

GUIDED EXAMPLES:

1. Solve by multiplying or dividing. Check your solution. $5x = 30$

2. Solve by multiplying or dividing. Check your solution. $-8x = 14$

3. Solve by multiplying or dividing. Check your solution. $\frac{x}{-9} = -2$

4. Solve by multiplying or dividing. Check your solution. $\frac{5x}{6} = -\frac{1}{3}$

Name: ______________________________ Date: ______________

Instructor: ___________________________ Section: ____________

Notebook 12.8
Solving Equations – Two Steps

OTHER NOTES

To solve an equation means to find the value or values that, when substituted into the equation in place of the variable, makes the equation ________ .

EXAMPLES 1–3 Solve the equation using the properties of equality.

$x+5=3$

$3x=-12$

$3x+4=10$

What is the procedure for solving equations using both properties of equality?

1.

2.

3.

4.

EXAMPLES 4–6 Solve the following equations. Check your solutions.

$4x-5=7$

$-2x+7=5$

$\frac{1}{2}x-4=-1$

EXAMPLES 7 & 8 Solve the following the equations.

$$\frac{x}{5}+2=-7$$

$$-4x+7=-3$$

EXAMPLES 9 & 10 Solve the following equations.

$$-4=1+5x$$

$$\frac{4}{5}=\frac{1}{5}+2x$$

CONCEPT CHECK:

Describe the steps used to solve $3x + 5 = 16$.

GUIDED EXAMPLES:

1. Solve. $3x-7=-1$

2. Solve. $-4x+2=6$

3. Solve. $7=3-5x$

4. Solve. $6.2x+5.3=-19.5$

Name: ______________________________ Date: ______________

Instructor: ___________________________ Section: ______________

Notebook 12.9
Solving Equations – Multiple Steps

OTHER NOTES

What is the procedure for solving equations with multiple steps?

1.

2.

3.

4.

5.

6.

EXAMPLES 1 Solve. $3(x+5)=-6$

Remove parentheses.
Simplify each side.
Get the variable terms on one side of the equation.
Get the number terms on the other side of the equation.
Get the variable alone on one side of the equation.
Simplify.
Check your solution.

EXAMPLE 2 Solve. $5x+4-3x=-8$

Remove parentheses.
Simplify each side.
Get the variable terms on one side of the equation.
Get the number terms on the other side of the equation.
Get the variable alone on one side of the equation.
Simplify.
Check your solution.

EXAMPLE 3 Solve. $6x+7=2x-5$

Remove parentheses.
Simplify each side.
Get the variable terms on one side of the equation.
Get the number terms on the other side of the equation.
Get the variable alone on one side of the equation.
Simplify.
Check your solution.

EXAMPLE 4 Solve. $5(2x-4)+7=3(x+5)$

Remove parentheses.
Simplify each side.
Get the variable terms on one side of the equation.
Get the number terms on the other side of the equation.
Get the variable alone on one side of the equation.
Simplify.
Check your solution.

CONCEPT CHECK:
List the seven steps that may be needed in solving a multiple step equation.

GUIDED EXAMPLES:

1. Solve $-9(x+1)-2=7$. Do the check on your own.
2. Solve $5x-6.4-x=-4.8$. Do the check on your own.
3. Solve $3(x+3)=x-3$. Do the check on your own.
4. Solve $3(2x-5)=7(x-3)+6$. Do the check on your own.
5. Solve $2+3(x-5)=-16$. Do the check on your own.